WindEurope Conference & Exhibition 2017

AA006789

Journal of Physics: Conference Series Volume 926

Amsterdam, The Netherlands
28 - 30 November 2017

Editors:

Michael Muskulus
Carlo L. Bottasso

ISBN: 978-1-5108-5273-0
ISSN: 1742-6588

Printed from e-media with permission by:

Curran Associates, Inc.
57 Morehouse Lane
Red Hook, NY 12571

proceedings.com

Some format issues inherent in the e-media version may also appear in this print version.

Printed by Curran Associates, Inc. (2017)

For permission requests, please contact the Institute of Physics
at the address below.

Institute of Physics
Dirac House, Temple Back
Bristol BS1 6BE UK

Phone: 44 1 17 929 7481
Fax: 44 1 17 920 0979

techtracking@iop.org

Additional copies of this publication are available from:

Curran Associates, Inc.
57 Morehouse Lane
Red Hook, NY 12571 USA
Phone: 845-758-0400
Fax: 845-758-2633
Email: curran@proceedings.com
Web: www.proceedings.com

Table of contents

Volume 926

WindEurope Conference & Exhibition 2017
28–30 November 2017, Amsterdam, The Netherlands

Accepted papers received: 2 November 2017
Published online: 23 November 2017

Preface

Health & safety on land and at sea

Resource Assessment (on- and offshore)

Supply chain, logistics and O&M on land and at sea

IOP Conf. Series: Journal of Physics: Conf. Series **926** (2017) 011001 doi:10.1088/1742-6596/926/1/011001

Preface

On behalf of the European Academy of Wind Energy (EAWE[1]) and WindEurope (formerly the European Wind Energy Association EWEA), we hereby have the pleasure of presenting the Scientific Proceedings of the WindEurope Conference & Exhibition 2017.

WindEurope Conference & Exhibition 2017 was organized in cooperation with the Netherlands Wind Energy Association (NWEA) and took place in Amsterdam on 28-30 November 2017. It is one of the largest and most interesting meeting places of the wind energy community, featuring 35 thematic sessions with more than 120 presentations this year. Known previously as *EWEA's Annual Conference* and as the *European Wind Energy Conference* before that, it has included Science & Research topics since 2007. First organized as a separate Science & Research track, the scientific contributions are since 2015 integrated into the sessions together with the more industrially oriented presentations, in order to increase the visibility of relevant scientific research and to facilitate an interchange of ideas between industry and academia. The event is a well-known meeting place that has for many years allowed engineers and scientists – both already established as well as young graduate students – to present their latest findings to a mixed audience consisting of both researchers and industry experts. Industry has presented the latest commercial developments, business cases and visions for the future, while scientists demonstrated relevant, novel ideas and approaches to problems of current interest. Very often this has led to interesting discussions and cooperative efforts that will continue well beyond the conference boundaries.

In continuation of the successful agreement between WindEurope and IOP Publishing, the scientific contributions of the WindEurope Conference & Exhibition 2017 are published in the Journal of Physics: Conference Series, an Open Access journal managed by the Institute of Physics (IOP). All papers in this volume originate from scientific contributions presented in the oral sessions of WindEurope Conference & Exhibition 2017, and have been peer-reviewed in order to ensure strict, international standards, coordinated by the Scientific Proceedings Editors and the Scientific Topic Leaders from the EAWE.

The proceedings in this volume consist of 15 papers covering a wide spectrum of wind energy research. Topics include performance monitoring of wind turbines using SCADA data, condition monitoring of wind turbine generators, wind speed forecasting based on numerical weather models, a study of inter-annual wind speed variability in Ireland, an analysis of the first German offshore wind energy auction, work on bearing loads and remote sensing, among others, as well as a study of lifetime extension of wind turbines and an evaluation of the ice throw hazard.

As with previous events, the European Academy of Wind Energy is responsible for organizing the review process for scientific contributions, it has contributed to the development of the sessions, and has provided Scientific Topic Leaders and chairs for all sessions. All papers were peer-reviewed by a Scientific Committee, consisting of scientists from EAWE member institutes and their associates. We thank all authors for their willingness to take part in this procedure, including the short deadlines

[1] http://www.eawe.eu/

imposed by the tight publication schedule. This was especially challenging this year, with most of the abstract review taking place over the summer, with only a few weeks left for the review of the papers in October. We thank all reviewers for their engagement and entirely voluntary work, without which we could not guarantee the scientific quality and current interest of the accepted contributions. WindEurope is responsible for the organization and logistics of the conference, and we thank their highly professional staff and their associates for the excellent collaboration.

Michael Muskulus
Editor Scientific Proceedings WindEurope Conference & Exhibition 2017
Scientific Programme Coordinator WindEurope Conference & Exhibition 2017
Norwegian University of Science and Technology NTNU

Carlo L. Bottasso
Editor Scientific Proceedings WindEurope Conference & Exhibition 2017
President European Academy of Wind Energy EAWE
Technical University of Munich TUM

IOP Conf. Series: Journal of Physics: Conf. Series **926** (2017) 011002 doi:10.1088/1742-6596/926/1/011002

Peer review statement

All papers published in this volume of *Journal of Physics: Conference Series* have been peer reviewed through processes administered by the proceedings Editors. Reviews were conducted by expert referees to the professional and scientific standards expected of a proceedings journal published by IOP Publishing.

IOP Conf. Series: Journal of Physics: Conf. Series **926** (2017) 012001 doi:10.1088/1742-6596/926/1/012001

Understanding and acknowledging the ice throw hazard - consequences for regulatory frameworks, risk perception and risk communication

R. E. Bredesen[1], **M. Drapalik**[2], **B. Butt**[3]

[1] Kjeller Vindteknikk, Norway
[2] University of Natural Resources and Life Sciences, Vienna,
[3] The Norwegian Water Resource and Energy Directorate

rolv.bredesen@vindteknikk.no, markus.drapalik@boku.ac.at, bubu@nve.no

Abstract. This study attempts to provide the necessary framework required to make sufficiently informed decisions regarding the safety implications of ice throw. The framework elaborates on how to cope with uncertainties, and how to describe results in a meaningful and useful manner to decision makers. Moreover, it points out the moral, judicial and economical obligations of wind turbine owners such that they are able to minimize risk of ice throws as much as possible. Building on the strength of knowledge as well as accounting for uncertainty are also essential in enabling clear communication with stakeholders on the most important/critical/vital issues. With increasing empirical evidence, one can assign a higher confidence level on the expert opinions on safety. Findings regarding key uncertainties of ice risk assessments are presented here to support the ongoing IEA Wind Task 19's work on creating the international guidelines on ice risk assessment due in 2018 (Krenn et al. 2017)[1-6]. In addition the study also incorporates the findings of a Norwegian information project, which focuses on the ice throw hazard for the public (Bredesen, Flage, Butt, Winterwind 2018)[7-9]. This includes measures to reduce damage and hazard from wind turbines for the general public. Recent theory of risk assessment questions the use of risk criteria for achieving optimum risk reduction and favours the use of the ALARA (as low as reasonably achievable) principle. Given the several practical problems associated with the ALARA approach (e.g. judicial realization), a joint approach, which uses a minimum set of criteria as well as the obligation to meet ALARA is suggested (associated with acceptable cost). The actual decision about acceptance criteria or obligations is a societal one, thus suggestions can be made at best. Risk acceptance, risk perception and risk communication are inextricably linked and should thus never be considered separately. Risk communication can shape risk perception, which again is vital for defining risk acceptance. Moreover, risk communication should be seen as an opportunity to demonstrate trustworthiness and an open, responsible and caring attitude. It is important for the wind industry to avoid accidents: In Winterwind 2017 (Ronsten)[10], the importance for the wind power community to proactively take safety measures for passers-by and service personnel was emphasized: Establishing good practices and communication routines is key to avoid accidents. Visually attractive ways of presenting the risk of ice throw are recommended.

1. Introduction

All wind turbines operating in an icing climate introduce the hazard of ice throw or icefall on the nearest surrounding area although with highly varying frequency and probability. The associated risk can be mitigated, but choosing the best, sufficient or optimal methods requires a good understanding of this phenomenon, and other related risks relevant for a given planned or built wind farm. Knowing, -among others the ice risk magnitude in time and location with sufficient accuracy and precision is essential in choosing the best and adequate risk reducing efforts prioritized by e.g. risk acceptance criteria, cost-benefit considerations and (available) resources. Building on the strength of knowledge and accounting for uncertainty is also essential to enable clear communication with stakeholders on the most important issues.

Achieving nearly zero damages and fatalities from ice throw (Vision Zero) means that there is a public requirement of warning systems for the risk of ice throw. It has been a challenge to get the warning to be respected; therefore it is important to achieve good communication. The risk of ice throw is highest during icing and melting conditions but can also persist into good weather conditions. Also, the meteorological conditions at ground level may be different than in the rotor plane of the turbine. The most important point is that the owners must understand the risk picture in their own wind farms in order to clearly and effectively communicate with third parties and stakeholders.

Apart from the moral obligation, there is a judicial and economical responsibility, since criminal and compensatory liability depends on the commitment to perform effective risk mitigation efforts. This is tied to a personal responsibility for third-persons to respect signs and warning systems, assuming they are reasonable, i.e. well designed and only actively warn when there is a danger. However, the danger can be avoided. Moreover, risk communication should be seen as an opportunity to demonstrate trustworthiness and an open, responsible and caring attitude.

Additionally, different approaches to safety assessments are necessary for different demands: For questions of siting and regulations, a metric for comparison of the risk of (sometimes entirely) different systems is necessary. In this case the use of LIRA (Localized Individual Risk per Annum) contours with the established 40 J impact energy limit for fatal ice fragments should be applied. When including safety systems such as deicing, the reliability and effectiveness of these systems needs to be considered as well. It is estimated that dangerous ice amounts occur when the ice load on a reference object located at hub height reach a threshold in the vicinity of 0.5 kg/m, that is the border between Ice Class Rime 1 (ICR1) and Rime 2 (ICR2) following the ISO:12494 standard on atmospheric icing of structures [11] (see Figure 1).

Figure 1. Pictures presented by Wadham-Gagnon, M., (Winterwind2013)[12]. Ice profile Classification Based on ISO 12494. Courtesy of TechnoCentre éolien (Wind Energy TechnoCentre).

IOP Conf. Series: Journal of Physics: Conf. Series **926** (2017) 012001 doi:10.1088/1742-6596/926/1/012001

When addressing the placement of warnings (e.g. signs) or barriers (e.g. closing roads) only the relevant turbine with its specific surroundings needs to considered and the safety goal will be extended to avoiding serious harm to people. In this case either different parameters for the existing metric or other conservative rules (e.g. the Seifert distance rule of 1.5 *(hub height + rotor diameter)) may be used [13-14]. Additionally warnings and barriers must be placed in a way that following them does not overly inconvenience the affected.

While methods for estimating icing frequency and ice loads on wind turbines as well as ballistic models for prediction of ice throw or shed distances are fairly advanced, thorough risk assessment demands the inclusion of uncertainties. This is seriously hindered, since these models are intrinsically deterministic and thus not suitable to give uncertainties per se. However, the probabilistic nature of ice throw and icefall can be partly considered by accounting for turbulence in the employed trajectory models. Also, using experiments with dropped ice fragment replicas and comparison with monitoring data, these uncertainties can be estimated and thus boundaries set.

Although the knowledge/research about uncertainties in ice risk assessments is increasing, this is far from sufficient. Further studies on trajectories of thrown ice, the impact kinetic energies, and the amount and shape of ice thrown are therefore important for increasing the credibility of risk assessments. Finally, since it is impossible to avoid all accidents further planning for communication of this fact is necessary.

In this paper we make use of the openly available data of collected ice pieces from the Swedish IceThrower project [15-18] in addition to similar databases from 4 wind farms in Austria. These data are described in Chapter 2 [19-22]. This chapter also contains the experimental setup of a conducted ice throw experiment along with a description of the risk assessment tool, IceRisk, developed by Kjeller Vindteknikk [23-25]. In Chapter 3, the results from the ice throw experiment are presented and discussed. This experiment and the gathered observational data is compared with results from IceRisk. The observations and model results for the basis of a new formulation for throw distances from an idling turbine with a dual detection system is presented in Section 3.3. A discussion on accuracy and uncertainty of modeled ice throw is given in Chapter 4. Finally a description of the national Norwegian guidelines regarding risk of ice throw, risk communication and the legal consequences in Chapters 5-8. The most relevant resources not specifically referenced in this document are listed in the bibliography [37-39] and [40-70]. Especially important are the documents Recommended Practices (2011, 2017 [13]) and Available Technologies (2016) by the expert group on wind energy in cold climates: International Energy Agency-Wind RD&D Task 19, https://www.ieawind.org/task_19.html.

2. Methodology

This chapter describes the data sources, experimental setup and models used.

2.1 IceThrower observations

The Swedish IceThrower project [15] produced the first open database of collected ice pieces from several wind farms [16], a detailed project report [18] with an updated recommendation on the set-back distance for Sweden as well as publicly available numerical software for calculating and presenting probability-distance functions for site- and turbine-specific risk assessments [16]. Results from the project has been presented by Lunden (2017)[17] and compared with the IceRisk model in Drapalik and Bredesen (2017)[22]. The project report (Göransson et al, 2017)[18] suggest H+D (hub height + rotor diameter) as a setback distance for Sweden. Figure 2 shows the statistics from 8 different ice throw cases from one of the turbines in the project.

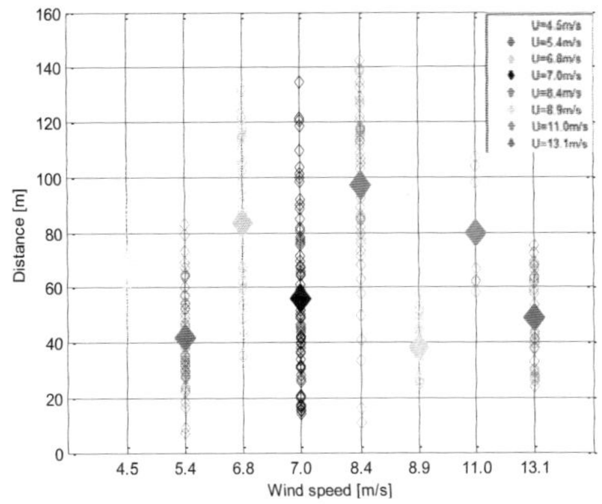

Figure 2. Statistics on collected ice pieces thrown from a V90 turbine in one wind farm in Sweden from 8 separate episodes (event A-H). Throw distance is given on the y-axis, and measured wind speed on the x-axis. (courtesy of Jenny Lunden, Pöyry, 2017)

IOP Conf. Series: Journal of Physics: Conf. Series **926** (2017) 012001 doi:10.1088/1742-6596/926/1/012001

2.2 Austrian field campaign for collecting ice debris

Legislation in Austria demands for turbines to be shut down when icing is detected, so only ice shed from still standing turbines can occur. In a monitoring campaign ice shed events have been observed and ice fragment positions and weights recorded [19]. Table 1 gives an overview of the conditions at the sites and the number of collected fragments. Detailed data of the monitoring program will be published as open online database on risk.boku.ac.at.

Table 1. Overview of monitored wind parks. Mean wind speed is the average wind during the recorded ice shed cases (double detection system of icing for stopping the turbine allowing idling).

Wind park	# of collected recorded fragments	Mean wind speed (1 h) [m/s]	Hub height [m]	Rotor diameter [m]
WP 1	32	12	138	82
WP 2	182	6	135	101
WP 3	231	6	113	71
WP 4	34	4	65	70

2.3 Experimental ice throw setup

In the course of the development of new methods for the assessment of small wind turbines (SWT) for urban use an experimental investigation of ice throw from SWTs was carried out [19,20]. In a short monitoring campaign several thrown ice pieces were collected and 3D scanned. These scans were used to produce realistic specimens via 3D printing. A throwing device was constructed with 1 m rotor radius and a maximum rotational speed of 200 rpm. The specimens were accelerated to the desired rotational speed and released at a defined rotor position. The throw trajectories were recorded with 3D imaging techniques.

2.4 IceRisk

In the IceRisk model, hourly time series of wind and icing conditions from a long-term meteorological simulation is used as input to the ballistic model described by Biswas [26]. This simplified ballistic model is considered valid for calculating the average (deterministic) trajectories associated with a given site-specific wind profile and the shape/form and density/weight specification of relatively compact ice pieces. That is, when the effects of the varying drag and lift, auto-rotation, and turbulence can be neglected for the considered ice pieces. For a site and turbine specific analysis where the purpose is to derive statistics on how the ice throw and ice fall hazard varies with distance and direction from the construction, this procedure is justified if the relevant statistics no longer are sensitive to the considered simplification; e.g. the sample size is sufficiently large and the associated parameters are adequately discretized.

For now-casting and fore-casting purposes considerations regarding the location-specific turbulence can be made to resample instantaneous wind measurements or hourly forecasted wind speeds on a short time horizon. By applying the IceRisk model and accounting for turbulence intensity a map giving risk zones and ice fall/throw probabilities can be created. Examples of this is shown in Figure 3 for the 209 m tall guyed telecom mast located at Tryvann in Oslo. For the forecasted wind conditions, a risk zone shown to the left are spanned by employing the 40 J distance function (Bredesen and Refsum, IWAIS2015)[23] for average hourly values as shown to the left in the Figure (verified as highly realistic impact locations by field studies at this site). Now, also the ISO-risk contours associated with a factor 10, and 100 lower risk than given at the edge of the indicated risk zone (left) can be presented. The strongest winds before resampling is 14.4 m/s at 209 m resulting in a maximum 40-J distance of 127 m. The strongest gusts after resampling are 19 m/s from the north showing the probability level for a possibly fatal strikes out to 220 m with the given assumptions. Similar methodologies could also indicate 3 consequence regions. e.g. of worst consequence/hazard, serious injury, and nuisance based on the considered energy threshold.

The example resampling technique that includes variation in the wind speeed and wind direction within 10 minute intervals is: Hourly values of a 48 hour forecast are linearly interpolated to 10 minute values, then normal distributed for a width of 3 standard deviations for both the wind speed and direction. We have chosen a standard deviation value of 10 % on the wind speed (turbulence intensity TI = 0.1) and 20 degrees for the wind direction to illustrate the effect. Suitable values can be found by analyzing the wind measurements from the newly erect official 30 met mast at the location (Tryvann, Oslo) employing relations between maximum wind speeds and average wind speeds as described by Harstveit (1996)[27]. Also the height dependence on the variation of the turbulence should be considered.

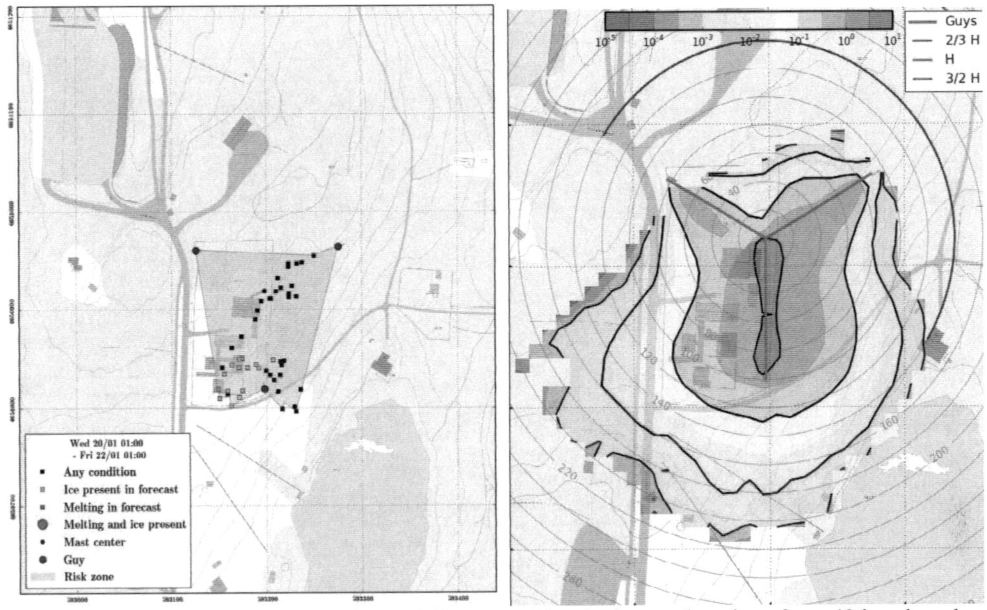

Figure 3. Left: Forecasted risk zone (ice fall) spanned by the impact locations from 48 hourly values of wind speed and direction using the 40 J distance function for ice debris released from a single point in top of the 209 m telecom mast. Right: Zoomed forecast of hazard zones with ISO-risk contours in multiples of factor 10. The probability map are calculated using the long-term size-distribution and resampled wind conditions, combined for all weighted release positions in the construction which consists of mast and 12 guys. Terrain effects have not been considered in the presented graphic. The indicated colored risk values are not scaled that is the absolute values are not representative to the risk at the given time, but the risk decay with distance is realistic.

Considering ice throw from the blade of wind turbines, results with the current setup are presented as the average of 4 separate calculations: 50 % by the wind statistics during melting conditions and 50 % by dangerous ice amount present on the blade (~instrumental icing). Each of these cases are then again subdivided by using an upper rpm-curve and lower reduced rpm-curve due to icing describing the velocity of the wind turbine blade as function of the wind speed. Figure 5 shows the average of these 4 scenarios for the considered V90 turbine in Sweden. In the IceRisk modeling, ice continues to accrete in repeated cycles during meteorological icing conditions. The maximum expected throwing distances of ice debris are associated with the form-factor (Cd A/M=0.008) solving the Biswas equations. Meteorological icing conditions are included as part of the period with ice present, but has not been weighted further. This is a natural improvement in the modeling as repeated cycles of shedding only occur during meteorological icing conditions.

3. Results

3.1 Dissemination of IceThrower results using the IceRisk Model

In the IceThrower field study 75 % of the ice was found within one rotor diameter distance from the turbine tower, and 1 % beyond 1.5 rotor diameter distance. 40 % of the ice debris weighed more than 1 kg and 3 % weighed more than 2 kg. The observed peak probability of recorded non-crushed ice debris in the IceThrower database occurs from 0-35 m distance with a probability at 8x10^-5 ice strikes/ ice event / square meter. The probability of ice debris impacting one square meter per event drops linearly with distance with 2 decades reduction over the next 100 m as shown in Figure 4.

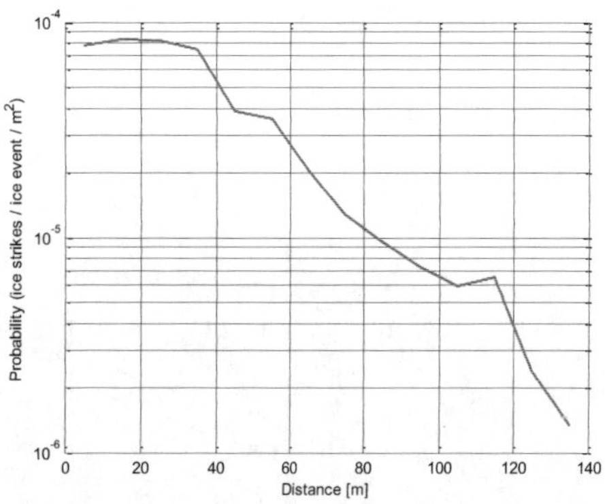

Figure 4. The probability of ice throw per m^2 and ice event based on all available data (532 pieces) from the field study. Courtesy of Bengt Göransson (Pöyry).

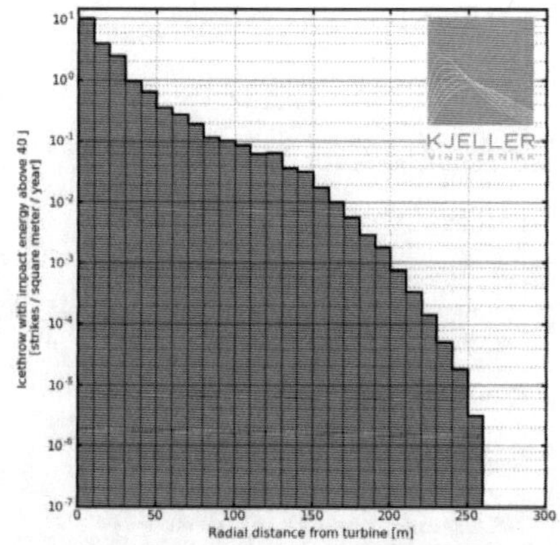

Figure 5. Results from an IceRisk-simulation which includes the calculation of the ice accretion on a turbine blade by Kjeller Vindteknikk (performance curves of a Vestas V90IIA turbine and a wind shear power law coefficient, alpha, of 0.21).

The results using the IceRisk method and the IceThrower software (Lunden, 2017) for the considered location in Sweden was compared in Drapalik and Bredesen (WWEC2017) showing very similar shapes on the expected radial probability distribution of ice throw. This discussion is a continuation of the study in Drapalik and Bredesen (2017, WWEC)[22]. The IceRisk simulation shown in Figure 5 is based on a long-term meteorological simulation [22, 23, 24]. Kjeller Vindteknikk recommends risk mitigation for ski-tracks and foot-paths especially within the distance associated with the millennium ice piece (with an impact kinetic energy level above 40 J), here calculated at 200 m.

An extrapolation factor based on measurements indicates that the associated LIRA (Localized Individual Risk) contours for power ten reductions of risk is at 250 and 300 m valid for turbines in comparable wind and icing climate (mid to north Sweden). Such a consideration is conservative as the risk decayed even faster with distance in the numerical simulations with rates of three decades reduction per 100 m.

By the use of the open Swedish IceThrower database on collected ice debris [16], a cross comparison is ongoing with meteorological modeled parameters as well as SCADA (Supervisory Control And Data Acquisition) data. This ongoing comparison is essential in increasing the precision of ice risk analysis in general, and to adjust the lower and upper uncertainty band as previously discussed in Drapalik and Bredesen (2017) [22]. Further results of this study will be presented in Bredesen (2017, Revision 1) [24] as a verification study of the IceRisk model.

In Figure 6 all collected ice pieces for events A-H from IceThrower are plotted by location (each location has been rotated to correct for the given wind direction for easy comparison.). From the collected ice pieces in the IceThrower project we observe that the maximum across-wind distance component is 120 m, while 140 m is the maximum throwing distance. We observe that directional wind-shifts have most likely occurred for at least two episodes: grey (13.1 m/s) and red (5.4 m/s).

For the 8.4 m/s event (purple) two 2 kg pieces were thrown 130 m for angles around 20 degrees from the along wind direction. Ice pieces weighing 2 kg were expected for higher across-wind to along-wind distances only (see Figures in Drapalik and Bredesen, 2017)[22]. Further comparisons, refinement and/or choice in modeling are therefore required to enable correct modeling of angular distributions.

Regarding the uncertainty on the absolute value of risk/probability we can make the following simplified comparison: from the 8.4 m/s episode (event E, purple markers), we count for example 20 impacts in a 40 x 40 square meter area between 100 and 140 m for a 40 m wide area at angles below 45 degrees. This gives an average areal strike probability of 0.0125 strikes per square meter. For comparison the IceRisk simulation result (Figure 5) was 0.06 strikes/square meter per year at a distance of 120 m which includes crushed ice. A similar exercise can be performed for example for the 4.5 m/s episode (A: yellow markers). This simple comparison indicates that the IceRisk simulation gives realistic results without being overly conservative in terms of the absolute risk/probability level for the indicated area. However, the risk is lower for other directions relative to the turbine location resulting in the lower observed directional averaged probability density of non-crushed ice debris reported in Figure 4.

A preliminary comparison of the SCADA data in terms of power production indicates only minor penalties on the performance (short-term) for most of the ice collection episodes, that is the ice amounts associated with the collected ice debris did not stop the operation of the turbine. The turbine was in operation at all ice events except for 11 December 2013 when the turbine was manually paused (event G, blue, 11 m/s).

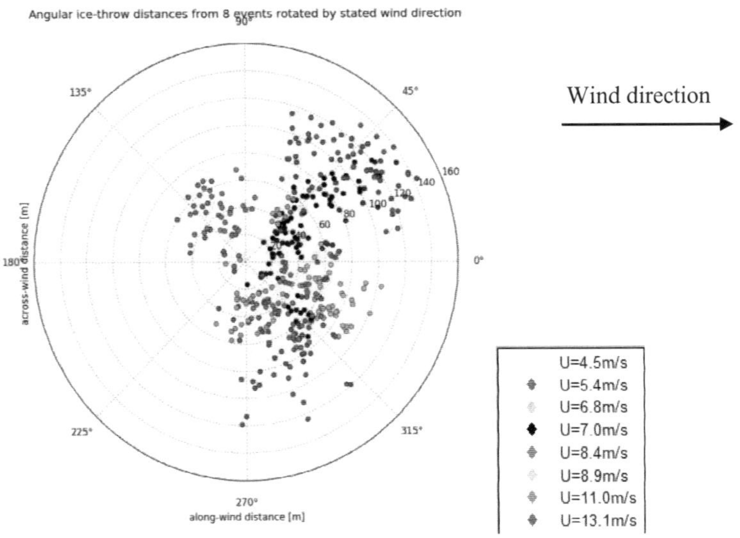

Figure 6. 417 ice pieces from the IceThrower database for the considered V90 turbine with a tipheight of 140 m. The location of all ice pieces are rotated by the given wind direction for each given case. Events are listed A-H by colored markers for increasing wind velocities.

3.2 Results from the ice throw experiments

Comparing the first results from the experimental ice throw project with model calculations based on the Biswas equations [26] two main points need to be addressed: Firstly, the Biswas equations tend to overestimate the throw distance by 10-15% (see example in Figure 7). A closer investigation of the video data from the throws shows that after approximately half the traveling distance chaotic autorotation of the fragment starts. This could lead to additional drag resistance due to turbulent flow, which results in shorter distances.

Secondly, the distance traveled along the wind is largely underestimated by the simulation, as can be seen in Figure 8. While this distance is small compared to the total distance, it has a significant impact on a non-isotropic risk map. The results presented in Figure 6 indicate the same differences when compared to the model. This is discussed in more detail in Section 4.

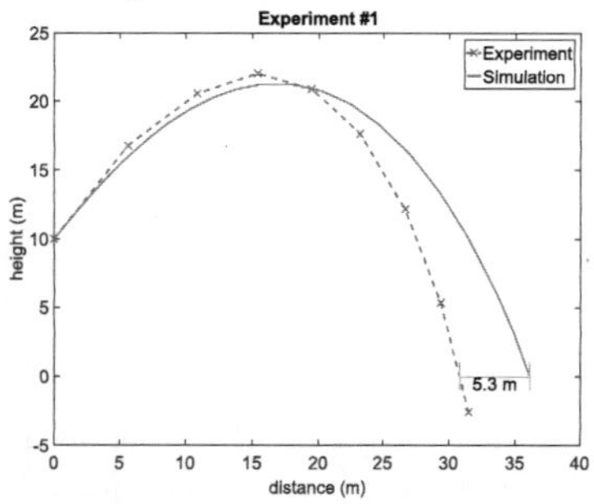

Figure 7. Comparison of calculated and experimentally found throw distance for 1 m radius, 200 rpm, 45° release angle and 4 m/s wind speed.

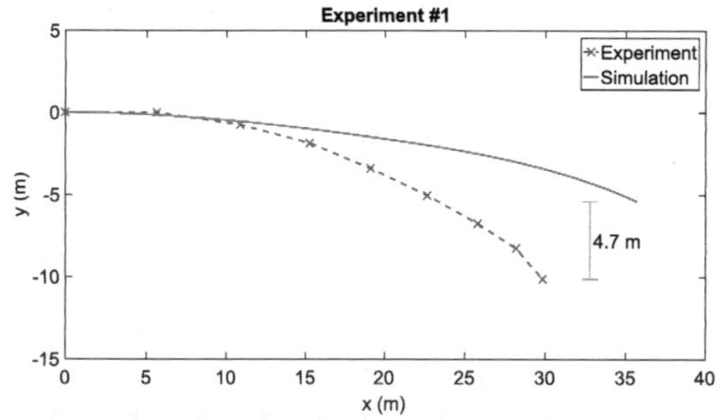

Figure 8. Comparison of calculated and experimentally found travel distance along wind for 1 m radius, 200 rpm, 45° vertical release angle along the positive x-direction and 4 m/s wind speed. The wind velocity is along the negative y-direction

3.3 Description of the Austrian ice shed samples and estimation of shed distances

For this Austrian field campaign the turbines were stopped (allowed to idle) during icing events. Figure 9 and 10 shows the distance distribution of collected fragments as well as the rotor radius and the expected maximum shed distance (d) from the Seifert formula (2003)[14]: $D_{max} = v \cdot (R + H)/15$, which is by far lower than the recommended 1.5 x total height by ISO12494 [11], which would correspond to the Seifert formula at 22.5 m/s. Here, v is the velocity at hub height (H) and R is the rotor radius. As expected, for low wind speeds and low hub height, most fragments drop within the rotor radius. In the case of the highest wind speed 55% of fragments were shed farther than the Seifert distance; the maximum distance found was 1.25 times the total height.

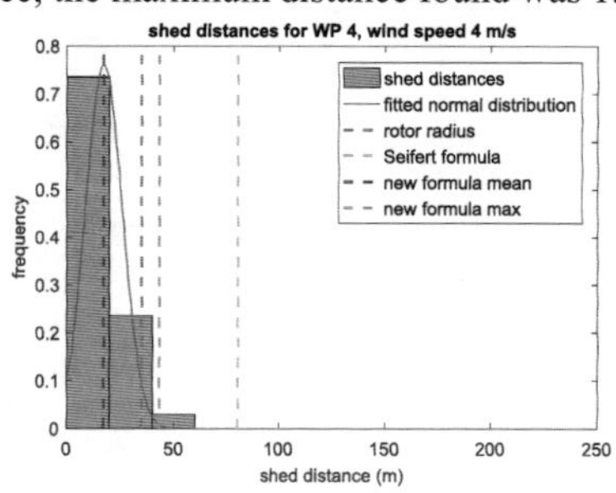

Figure 9. Shed distances for 4 m/s mean wind speed, 65 m hub height and 70 m rotor diameter

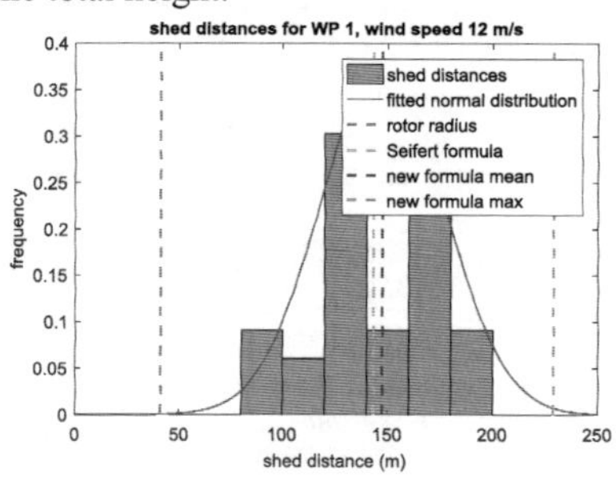

Figure 10. Shed distances for 12 m/s mean wind speed, 138 m hub height and 82 m rotor diameter. Outliers are included in shown bins.

For the high wind speed case in Figure 10 the highest shed distance was 225 m for a 240 g piece of glace ice. The thickness of this piece varied roughly from 3 to 8 mm as can be seen in Figure 11. Further pieces were found at distances around 150 m. For ice shed we note that thin and less compact ice pieces drifting the furthest with the wind have a lower energy density than more compact object reaching the same distance by ice throw. Also the damage potential considering a random orientation is lower compared to the most unfavorable setup possible.

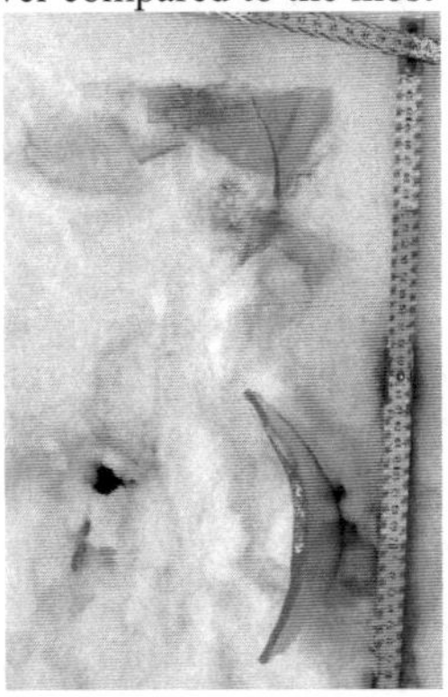

Figure 11. Ice piece with the highest recorded shed distance of 225 m from a turbine with 138 m hub height and 41 m rotor radius. The length of this piece was 38 cm, the width 16 cm and the thickness varies between 3 and 8 mm. It was pure glace ice with a weight of 240 g.

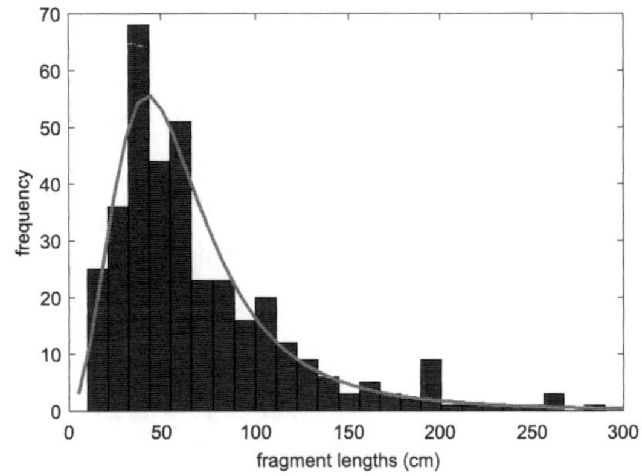

Figure 12. Histogram of the length of found ice pieces. A loglogistic function with $\mu = 4.04$ and $\sigma = 0.37$ is fitted to the distribution.

Size and geometry of the found ice pieces varied strongly and ranged from 20 cm to 3 m. Most fragments found had a length between 40 and 60 cm. Figure 12 shows a histogram of the found lengths with a fitted log-logistic function. For several of the observed events a positive correlation between length and width of the ice pieces was found as well as increasing fall distance with decreasing fragment size. This is expected as lighter and less compact fragments travel further. All of the largest ice pieces was found for distances shorter than H. A scatter showing the distance and size of the ice pieces is shown in Figure 13.

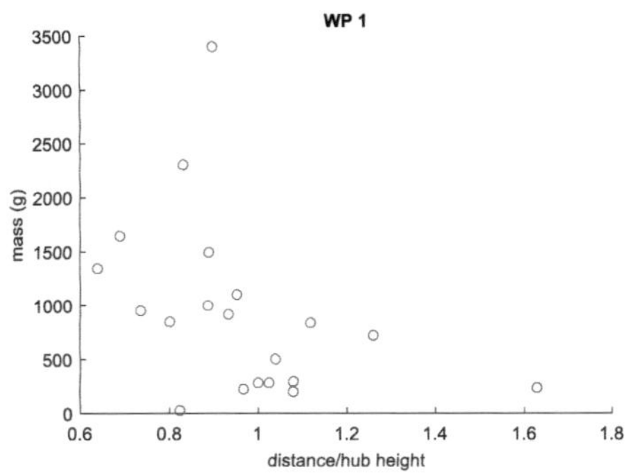

Figure 13. Scatter plot of mass and distance of found ice fragments in WP 1. The shed distance decreases with increasing mass, but only for high distances.

When trying to generate an empiric formula for maximum shed distances, a simple assumption can be made: the mean drop distance needs to reduce to zero for zero wind speed as well as for zero drop height. Next, it is assumed that drop distance increases linearly with the two mentioned factors. Finally it can be reasoned, that the mean drop distance should depend on the hub height instead of the total height, since the rotor diameter should only affect the maximum distance but not the mean. Applying this to the data from the four recorded events the following expression for the mean drop distance can be found:

$$D_{mean} = \frac{v \cdot H}{11.9} \tag{1}$$

To get the maximum distance, it is here assumed that the fall distances are normal distributed (this assumption is not neccessarily true and can be improved on especially with emerging evidence regarding higher wind speed and longer observed drifting distances, e.g. including results from Drapalik and Bredesen (2017) [22]). The width of this distribution depends on wind speed and total height, resulting in the following formula for the relative width of the fall distance distribution:

$$w_{rel} = 0.57 - \frac{v \cdot (H+R)}{5570} \tag{2}$$

Defining the maximum shed distance with the 3 sigma limit (99.9% of fragments are found within this limit), the maximum distance becomes

$$D_{max} = D_{mean} \cdot (1 + 3 \cdot w_{rel}) \tag{3}$$

While this estimation is by far more complex than the Seifert formula, it is entirely based on monitored data and can still be easily calculated for a given turbine.

Since these equations are modeled on relatively few monitoring cases it is limited to the range of these samples. For practical use we recommend the use for up to 15 m/s wind speed, 150 m hub height and a rotor radius to hub height ratio smaller than 0.5. Low wind speeds and large rotor diameters may also be problematic, since the equations are based on the assumption of normal distributed distances. This is based on the assumption that the distance along the wind direction will be more important than the distribution across the wind direction resulting from the rotor radius and the yaw and roll angle.

Figure 14 shows distributions of the monitored data scaled linearly for wind speed and hub height, with the estimated mean distance and maximum distance from equations 1 and 3. The maximum and mean distance for two of the pieces are also shown in Figure 9,10.

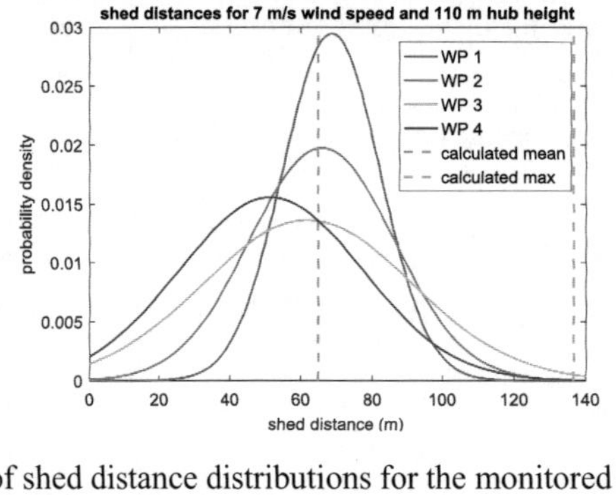

Figure 14. Comparison of shed distance distributions for the monitored wind parks. Distances have been modified with linear factors to reach the target wind speed and hub height. The dashed lines signify the calculated mean distance from equation 1 and the maximum distance from equation 3, assuming a 50 m rotor radius.

4. Discussion on uncertainties

For icethrow calculations, further comparisons, refinement and/or choice in modeling are recommend to enable better modeling of the angular distributions of ice throw. As a starting point it is recommended to include the effect of turbulence, the rotor tilt angle and to adjust the average drag coefficient from 1.0 towards 1.2 in the model. For non-compact debris shapes, the modeled ice shed drift distances are underestimated with the current assumptions as demonstrated in Figure 8 and 10.

No risk assessment should be based solely on the collection on uncrushed ice pieces found without considering the total amount of ice being released from the construction during an icing event. The amount of crushed ice debris in terms of number of throws can sometimes be visible in the form of craters on the ground if a snow cover exist. It is important to acknowledge that larger ice pieces are more likely to crush on impact which can cause a systematic bias in ice risk assessments. The IceRisk results presented in Figure 5 avoids this specific uncertainty by modeling the ice accretions directly on a representative model of the wind turbine's blades.

Ice risk assessments are especially sensitive on the operational mode of a turbine. During icing conditions, some turbines shut down early either intentionally, due to the control system or performance degradation. Other turbines, on the other hand, can operate with ice on the blade permitting repeated cycles of re-accretions and shedding for even a single icing episode. Following the initial risk comparison made by Drapalik and Bredesen (2017), a cross-comparison of the considered turbine's SCADA data, long-term simulations of meteorological parameters, and cataloged ice debris data collected from 8 separate icing episodes is ongoing. Furthermore, the timing of field observations to modeled and observed icing events are inspected as well as the performance degradation of the turbine during icing periods. This comparison is essential in increasing the precision of ice risk analysis in general, and to adjust the lower and upper uncertainty band as previously discussed in Drapalik and Bredesen (2017).

Most of the ice pieces collected in the IceThrower database come from work by Dala Vind in one of their wind farms consisting of V90 turbines. When their wind farm was built in 2008, a safety distance of 160 m was made in agreement between operators, responsible stakeholders of a surrounding ski track, and the wind farm service technicians. Here, 160 m is also the minimum distance between the ski track and one of the turbines in the wind farm. The ski track grooming crew performed ice collection in Dala Vind's wind farm on behalf of the IceThrower project. The wind farm is located in forested terrain, and the search has been thorough in the surroundings independent on search direction, distance and vegetation. For this location is not easier to perform the search at 120-140 distance compared to e.g. distances of 140-170m. For the 8.4 m/s episode, it was coincidence that most of the ice debris landed in the one-sector with more open terrain. As a consequence of the field study an updated recommendation of a safety distance of H+D (=185 m) is given the Swedish Report. Also worth noticing is that ice debris has not been observed on the mentioned ski-track by the grooming crew. H+D for this wind farm is 90+95 = 185 m. The furthest observed ice piece thrown from this wind farm was within the tip height of the construction (140 m). Based on the preliminary comparison performed here it is reasonable to assume that the IceRisk results with the current model setup (exlucing [23]) are conservative by a factor of 5 accounting for the number of ice pieces being thrown. As a result of this consideration the millennium ice piece distance (IceRisk) presented in figure 5 is no longer at 200 m but expected within the 175 m distance for the considered location.

4.1. Discussion of risk decay with distance

In Figure 4 one can note that for distanced beyond 40 m there is approximately a decadal reduction in probability for every 50 m in the range 40-140 m where ice pieces were found. If an assumption of 50 m decadal decay risk outside 40 m is made we can present the fraction of thrown pieces that are expected outside given distance circles. This is useful since large distances are not represented in typical field studies due to natural limitations in the large search areas and the lower probabilities. With the employed function 1 out of 1900 ice pieces will be expected outside the H+D distance, that is less than 0.1 % of the ice pieces. In addition 1 out of 100 000 ice pieces are expected outside the

Seifert distance of 1.5*(H+D). For a distance of 350 m from the turbine 1 out of 4 million ice pieces are then expected (see figure 15). Since there are no field studies that have found ice pieces at these distances and the probability of a field study finding ice at these distances is so low, a top down evaluation for Sweden was made. Assuming average installed capacity during the years 2012-2016 of 5228MW, all turbines were of the V90-2MW type studied, and all turbines are exposed to light icing thereby casting 200 dangerous ice pieces per year. This means total thrown dangerous ice pieces in Sweden during these five years exceeds 2.6 million, 1400 should have been cast further than H+D (185 m), and 26 000 beyond the tip height (140m). Because of their rarity, ice pieces found at long distances are expected to be points of discussion among neighbors, wind farm operators, service personnel, and media even when found outside of planned field studies. The number of reported ice pieces found beyond tip height in the media and among operators seems to be too low to support an extrapolation of the decadal decay beyond the range where ice pieces were recorded in the Icethrower study, and even more so if it is considered the total number of turbines in Europe which are exposed to icing conditions. Independent of the ice thrower study the author knows about only one far throw or fall (third hand anecdotal evidence); that is of one fist size ice piece (glaze) that has been observed at a distance of 300 m +- 50 m by a service technician for a site in south-eastern Europe with a different wind and icing climate than in Sweden.

To built on the empirical evidence we note that Enercon for 3 years has collected 25 000 ice pieces from 4 E82 turbines with different operating modes in one of their wind farms (as presented by Héctor Rodríguez et al., 2017) [3]. The maximum observed distance is there 170 m, which is adjusted to 203 m accounting for uncertainty. Presented ice throw and ice fall maps (simulations) show direction dependence with higher probability along the dominant wind directions.

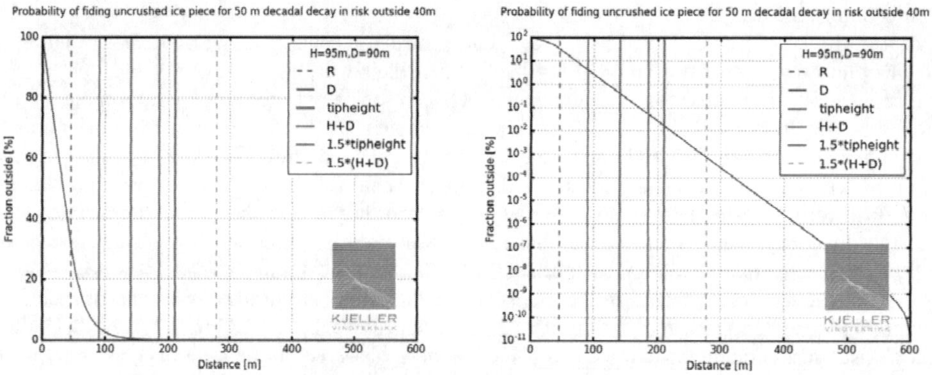

Figure 15. Fraction [%] (y-axis) of ice pieces thrown outside given distances (x-axis). The extrapolation function is based on the observed risk decay and presented on a linear scale (left) and logarithmic scale (right). Outside 200 m, the expected risk decay is even steeper with distance for this location (not shown).

4.2. Sensitivity study of ice piece size and wind speed distribution using the IceThrower model

The numerical software for the IceThrower project has been prepared by Jenny Lunden (Pöyry) available to the public valid for calculating omni-directinal strike probabilities given the wind conditions, turbine characteristics and size distribution of expected ice debris.

Pöyry's sensitivity study (Figure 16 and 17) showed that at distances up to 140 m (tipheight) the results were similar when 1) changing the average mass in the size distribution from 0.5 kg to 3 kg, and 2) for the combination of increasing the wind speed average from 7.6 m/s to 9.6 m/s and the mean mass of fragments from 0.5kg to 2 kg (for 100 000 modeled throws). It shows also that the probability of ice pieces at distances further than tip height is highly dependent on the mass of the ice pieces assumed near the blade tip where the highest velocities are found. Figure 16 shows a 1 decade difference in probability at H+D (185m) and the difference increases to a two decade difference in probability at 225m. Observations indicate that the largest ice pieces are found near the middle of the blade where the ice forms on a larger leading edge radius and there is less flexing of the blade and lower aerodynamic forces acting to break loose the ice (Homola, p. 6) [71]. Ice accretes fastest near the tip, but is broken off with limited thickness (observations seem to show about 6-7cm) when the forces acting on the ice exceed the ice strength. This effect, which limits the maximum size of ice accretions near the blade tip, could explain the relative lack of reports of ice pieces at distances beyond 140 m. This illustrates the sensitivity in the modeling of ice throw for wind farms. It will be important also to communicate such uncertainties in an ice risk assessment.

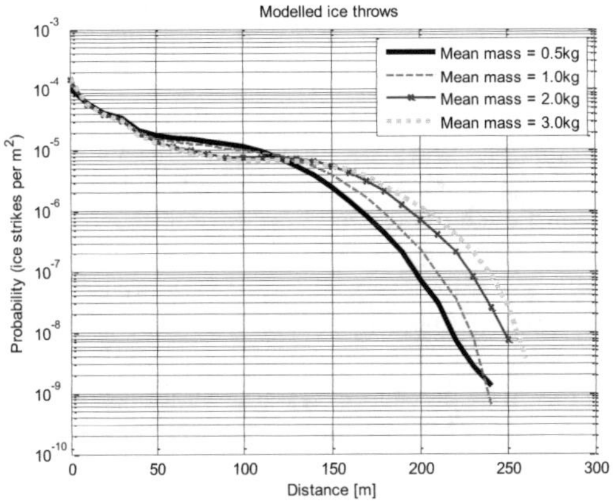

Figure 16. Probability, ice strikes/ice event/m2, for four random mass distributions with mean values of 0,5, 1,0 2,0 and 3,0 kg. The random wind distribution has a mean wind speed of 7,6 m/s. The turbine rotor radius = 45 m and hub height = 95 m. Courtesy of Bengt Göransson (Pöyry)

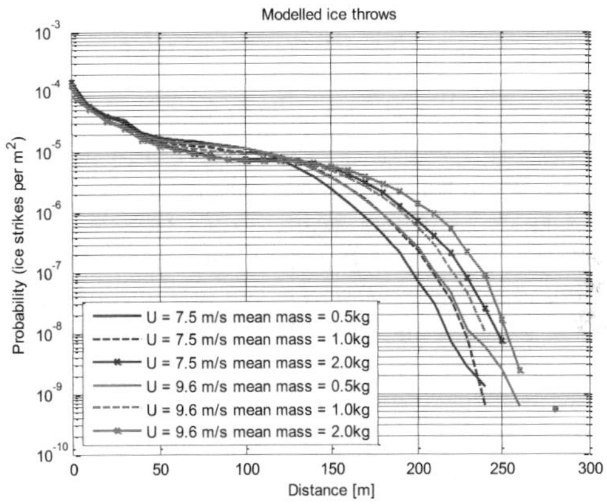

Figure 17. Probability, ice strikes/ice event/m2, for distribution with mean wind speed at hub height of 7,6 m/s (blue) and 9,6 m/s (red) and for three distributions of mass (average of 0,5, 1,0 and 2,0 kg). The turbine used has a rotor radius =of 45 m and 95 m hub height = 95 m. Courtesy of Bengt Göransson (Pöyry).

5. Risk and Risk communication

The classical definition of risk – risk = probability x consequences – is effectively tied to three assumptions.

First, the consequences in questions must acceptable regardless of the probability. In other words: some consequences are in any case inacceptable and must not occur.

Second, probabilities and consequences are quantifiable and the knowledge about them is sufficient. Quantification of consequences is sometimes difficult due to practical or ethical reasons, but needs to be made in a way that allows comparison with acceptance criteria. Quantification also demands for a test of the strength of knowledge, i.e. a thorough description of the necessary assumptions, simplifications and uncertainties. Methods of quantification that do not produce quantified uncertainty ranges for the used values are especially problematic in this context.

Third: The knowledge about possible events is sufficient. This can never be formally shown, but it is advisable to put reasonable effort in identifying events, increasingly so for complex systems e.g. chemical plants.

In communicating risks, these three assumptions should be explicitly addressed and the state of the assessment concerning them made clear.

5.1 Risk acceptance criteria

Risk acceptance criteria are always a result of an explicit (approval of a proposed regulation) or implicit (acceptance of established practices) societal decision.

Many existing criteria or principles are based on the comparison of a new risk with existing, similar risks. This is not applicable for the risk of ice throw, since no comparable situations exist. Thus one possibility is the application of the MEM (minimum endogenous mortality) principle, which demands that a single technical system does not increase the endogenous mortality by more than 5%. The endogenous mortality is usually chosen with 2×10^{-4} (mortality of a 15 year-old) [33], leading to a maximum risk of 10^{-5} deaths per year.

While this fixed value is convenient for practical use, it has several drawbacks. First, the choice of 5% is arbitrary and cannot be well communicated. Second, fixed values give no incentive for improvement of safety measures once this value has been reached. Especially the second argument makes the ALARA (as low as reasonably achievable) principle a considerably more attractive choice. The ALARA principle uses two limit values: the upper value defines an intolerable risk and demands like MEM that a risk is reduced until it falls below this value. The second limit value defines an negligible risk, which does not need to be reduced. The region between these values defines tolerable risks which need to be reduced a far as possible, i.e. until the costs are too high or until they become negligible. This definition incorporates a legal in addition to the moral obligation to reach the highest possible safety standards. While this concept principally supports far reaching and cost effective risk reduction, it is prone to misuse due to its rather open definition. The boundary condition of "reasonably achievable" is mainly dependent on the point of view of the user, leading to different understanding e.g. of operators and public. This approach should be openly communicated and the criteria for the choice of certain measurements made clear. The MEM principle is further described in the German study by Hahm and Stoffels (2016) [34].

5.2 Risk perception and communication

Risk acceptance, risk perception and risk communication are inextricably linked and should thus never be considered separately. Risk communication can shape risk perception, which again is vital for defining risk acceptance.

Risk communication is classically seen in a one-way direction – experts inform the concerned public of the risks and correct behavior, often after the completion of the associated project. This may result in additional opposition and little understanding. Additionally the risks of complex technologies are often difficult to explain lacking the relevant technological knowledge.

Thus, an early involvement of the public in project realization and application of safety measures is highly advisable.

6. The wind energy community's preferred policies regarding ice risk assessments

As there is a lack of understanding of the ice throw phenomenon among authorities, this has led to the fact that there is no coherence in the applied ice throw mitigation policies in various countries and regions, which can cause safety- and financial hazards for wind farms in icing climates (Nöel de Wild, 2017)[35]: "*Together with six researchers on ice throw from wind turbines, mitigation policy scenarios have been suggested in two forms:*

- *Method A, politically induced general regulations, like curtailment, obligated blade heating or safety zones around turbines.*
- *Method B, site specific mitigation methods based on guidelines for risk assessment.*

Method A is a clearly non-preferred scenario, however there is a fear of it becoming reality. Mandatory curtailment and safety zones could have a large negative impact on feasibility and development of wind power projects, whereas mandatory blade heating could give a false feeling of safety. Method B is therefore preferred. Having strict guidelines on how to perform risk assessment, will lead to site-appropriate mitigation solutions that are both safer, more understandable by the environment and less harmful to the feasibility of wind power development."

7. *National Norwegian guidelines* regarding the risk of ice throw for the public (2018)

The Norwegian guidelines [9] gives advice on the possible consequences of ice throw from wind turbines for the public (third persons). The guideline is the result of an internal project at the Norwegian Water Resource and Energy Directorate (NVE), with external contributions from Roger Flage at IRIS (International Research Institute of Stavanger), and Rolv E. Bredesen from Kjeller Vindteknikk.

The background for R&D project is that all wind farms in Norway to various degrees are exposed to icing conditions enabling the hazard ice throw and ice fall from the blades of wind turbines onto the surrounding areas. The resulting risk of injury is a topic for interested stakeholders and proper communication of this risk is a challenge, which may result in that the risk of being hit by ice debris originating from wind turbines sometimes are exaggerated. Therefore NVE has actuated the current Research and Development project regarding ice throw and ice fall which focuses on how the risk should be managed and communicated based on currently available documentation, best practices and research.

The goal of the project is to prepare easily understood and relevant advise for wind farm concessionaires (owners/operators) and to the general public:

1) How to communicate the risk of injury and damage to the general public caused by ice throw and ice fall.
2) Relevant measures handling the risk of injury and damage.
3) Clarify the criminal and compensatory liability for incidents involving injury.

The Norwegian Guideline is meant as a supplement to the IEA Wind Task 19's internationally harmonized guidelines regarding ice throw risk assessments. However, the health and safety issue related to wind farm working personnel is not considered in this project.

There is a requirement from NVE to evaluate the extent of icing and to assess the risk of ice throw/ice fall when applying for a wind farm license in Norway. An evaluation shall be sent to NVE before operating the wind-power stations.

A recommended list of the most important items in such an evaluation are presented together with short guidance on how to perform said work, including how to describe the related uncertainties and strength of knowledge regarding key assumptions in a risk assessment.

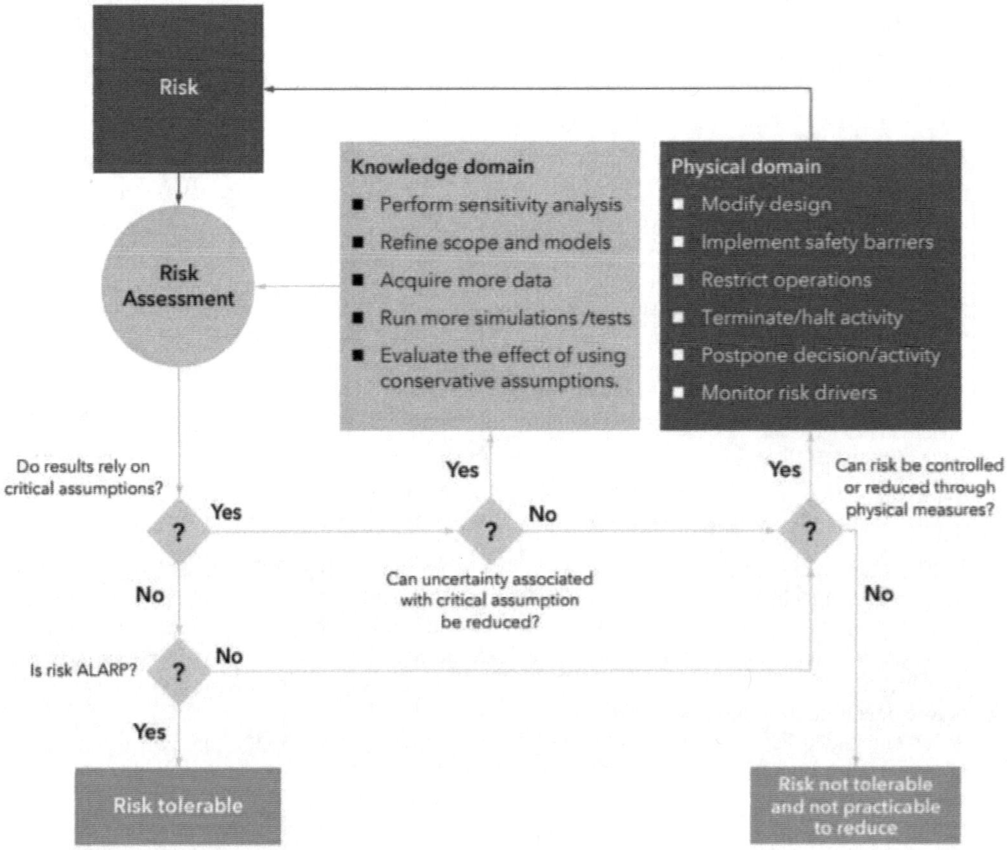

Figure 18. Risk assement as a tool to gain risk underderstanding, which is necessary to manage and keep the risk under control. The risk management process in the knowledge and physical domain enables sufficient understanding and confident decision-making. Source: DNV-GL (2016)[36] Enabling confidence: Adressing uncertainty in risk assessments. Courtesy of Frank Børre Pedersen (DNV-GL).

7.1 Risk assessment

Risk assessments are tools used to understand the risk related to an activity or a system in order to properly mitigate or accept the associated risks in a risk management process. One requirement regarding the quality of work and competence of risk analysts are on whether they are able to identify relevant risks and give concrete advice on the most beneficial risk reducing measures on a case by case basis. In any risk assessment it is important to appropriately assess and describe the strength-of-knowledge (e.g. the analyst's confidence in the presented results and/or underlying assumptions, models used, etc.) and the uncertainties involved (e.g. related to black swans (surprises) and completeness (aspects of the risk not described by risk indices/metrics)). Historically, risk acceptance criteria require a high accuracy of the risk assessment to be used as a sharp limit for when and where risk mitigation is required to consider the risk acceptable. The Norwegian National guidelines (2018) on how to perform ice throw and ice fall risk assessments acknowledges this precision issue, and accordingly question the procedure of using quantified risk estimates comparing calculated risk metric values against absolute risk acceptance criteria.

In its simplest form the required warning system (Norway and Sweden) for the ice throw hazard for the public is suitable located signs. This may be enough for remote areas that rarely are visited during the 'often' bad weather associated with icing conditions. It is therefore important for the risk assessment analyst to identify to whom, when (including how often) and where there is a risk of ice throw and ice fall to incorporate the best and adequate measures. Also if effective risk mitigation options are inexpensive to implement (cost-efficient) they should be implemented according to the ALARA/ALARP principle (As low as reasonably achievable/practicable associated with acceptable cost). The publicly available background knowledge regarding ice throw and ice fall is at the moment not necessarily strong but rapidly increasing. Key assumptions typically made in ice risk assessments are currently being scrutinized, thus improving the strength of knowledge underlying such assessments. The background knowledge related to ice throw and ice fall among experts is far from poor, and there is consensus among experts on where the large uncertainties lie.

The iterative process of gaining risk understanding in the knowledge domain and in the physical domain is shown in Figure 18. The key question is on whether the risk is as low as reasonably achievable.

7.2 Suitable risk mitigating measures

In addition to the detailed planning of the actual wind farms, designed signs and warning routines, possible sound and light signals, direct warning systems (e.g. sms/app), and detection systems it is recommended to establish public knowledge about the risk and to consider physical safeguards and/or curtailment options (e.g. to stop the turbine during periods of particularly high risk).

7.2.1. Establish knowledge about the risk

In order for the public to avoid injuries and damages due to the hazard ice throw and ice fall it is important that the threat is sufficiently understood leading to respect for the implemented signs and warning systems/routines. Signs and warning routines/systems should be supplemented with continuous and effective information transfer that helps the users of the area to understand the specific danger. This can be communicated through newspaper articles in local media, brochures / information in pocket format that can be distributed, QR codes on signs and more comprehensive information boards at the main entrance of the wind power plant. The information given should be repeated systematically prior to each winter season and reach different target groups. In some areas, it may be useful to provide information in several languages.

It is important that information about the risk is communicated in such a way that the receiver understands the message and becomes motivated to adapt his/her behavior. This requires a good dialogue and trust between the concessionaire and the users of the area. In areas of high usage, the concessionaire should facilitate involvement of users in assessing which measures are appropriate for managing the risk, so that stakeholder needs can be maintained through the actions that will be implemented.

It is an advantage that the information about the risk is correct and effective to ensure that the risk of damage from ice throw and icefall is not exaggerated or causes fear as this may cause the public to limit their stay in the area more than necessary for damage avoidance.

7.2.2. Physical safeguards and shutdown of turbines

In order to facilitate safe traffic and stay in the area, suitable measures can in some cases be different physical safeguards. If there are ski slopes, snowmobile routes and hiking trails, etc. in areas where ice throw and icefall may occur, these should be relocated and marked. It may also be appropriate with temporary cordons to prevent traffic during periods of particularly high risk of ice throw and ice fall. Other examples of security measures may be to reinforce the roof on the buildings located in exposed areas and roof over parking spaces, etc. to prevent damage to equipment and materials.

It may also be advisable to stop the turbines during periods of particularly high risk of damage from ice throw and icefall.

8. Legal consequences

Ice throw from wind turbines can potentially invoke serious legal consequences for wind farm owners in case the ice throw result in any serious damages. In Norway NVE can carry out supervision in both the construction and operational phases to ensure that the risk of damage by ice throws is handled in accordance with the current terms and conditions. In case of any discrepancy, NVE may react with different means, such as orders for rectification or termination of the license.

Terms and conditions for measures to deal with the risk of damage by ice throw are intended to prevent serious accidents. These accidents may occur even if the owners follow the regulations imposed by NVE. In such a scenario the affected party can claim compensation for any damages caused. The licensee of the wind farm can be held responsible even if all public-law requirements are satisfied. Owners and employees of the licensee may also be liable to criminal proceedings even if they act in accordance with public-law requirements set by the NVE. The police authorities investigate and prosecute criminal offenses.

8.1. Liability

The licensee may be held liable for damage caused by ice throw and icefall in accordance with the general principles of compensation. The affected party may claim compensation from whosoever he/she believes is responsible including board members, general managers and shareholders. In case the parties cannot arrive at a consensus, the case may be brought to the courts for the final decision.

There are three conditions that must be met in order for a liability to arise. These are discussed briefly below.

8.1.1. Basis for liability

To be held financially responsible for injuries the wind farm owners must have acted negligently. Both active actions and omissions may be classified as negligent and can therefore invoke the right for compensation.

In case law, the following factors are used to assess whether a party has acted negligently:

- The risk of injury is taken into account. The greater the risk is, the greater the requirement for attention. For example, if a wind farm is located in an area that is used extensively for outdoor activities, or if there is short distance to densely populated areas, more stringent requirements will be imposed on the attention than for wind farms in areas with limited human activity.

- What choice of actions the licensee had is also a critical factor. Emphasis is placed on whether the licensee could have avoided the damage by engaging in preemptive actions.

- Whether the licensee had complied with public-law rules is also a part of the assessment. However, the licensee may be held liable despite fulfilling all public-law requirements.

In addition to the basis of liability mentioned above, the owner of a wind power plant may be held liable without regard to negligence. According to Norwegian case law, there are three basic conditions that must be fulfilled in order to be liable without regard to negligencee: 1) There must be an extraordinary risk that significantly exceeds the daily life risks; 2) The risk of injury is typical for the current business. 3) The risk of injury must be persistent. For example, it has been established in Norwegian case law that the owners of apartment buildings are liable, without regard to negligence for damage caused by the collapse of parts of the building.[1]

[1] Rt. 1939 p. 766 (Gesimsdommen). Look also at Rt. 1972 p. 965 (Mønepannedommen)

IOP Conf. Series: Journal of Physics: Conf. Series **926** (2017) 012001 doi:10.1088/1742-6596/926/1/012001

8.1.2. Financial loss
The condition of financial loss will in most cases be fulfilled by damages to health and property.

8.1.3. Causation
There must be causation between the damage and the injurious act for the incident to trigger liability. In some cases, there may be multiple causes and more than one responsible for damage. An example might be where a tour operator facilitates traffic close to the wind farm without notifying the wind farm owners and as such increasing the risk of potential harm.

8.2. Criminal liability
According to section 10-5 of the Energy Act, violations of the Energy Act are punishable. It is the police, which investigates and prosecutes criminal offenses. The NVE may report an accident to the police. The police pursues violations of the provisions of the Criminal Code. Punishable circumstances can provide grounds for criminal proceedings against the enterprise, chairman, board members, operating company and employees / managers in the operating company. Supreme Court decision in Rt. 2013 p. 312 (Takrasdommen) is an example of criminal prosecution against both enterprises and individuals based on the Criminal Code after an ice cube of about 10 kilo fell down from the roof of an apartment building and hit a person. The Penal Code § 280 has a limit of imprisonment for up to 3 years for the person who "negligently causes significant harm to another's body or health." Penalty for causing death is up to 6 years.[2]

8.2.1. Corporate penalties
Penalties may be imposed on the enterprise in the form of a fine even if no one in the enterprise has been at fault. Penalties against the enterprise may be substituted for, or in addition to, penalties by individuals in the enterprise. In the example mentioned above about ice cube, the company was sentenced to a fine of NOK 2 million in the district court. This was adjusted to NOK 1.5 million in the Court of Appeal and in the Supreme Court the company was dismissed under dissent. In a case from 2009, both the company owning the apartment building and of the shareholder of the company were convicted of negligent killing after parts of a building fell down and caused death of a person passing by.[3]

8.2.2. Directors' liabilities
Board members of the company that own the wind farm may be liable to criminal proceedings for negligent harm. [4] It is the board's duties under the "*aksjeselskapsloven" (The Company law)*, which is the basis for the assessment. The board must ensure proper organization of the company and guidelines for the work.[5]

8.2.3. Operations manager
In cases where the cause of damage is due to negligent omissions or actions by the operating company or employees of the operating company, it may become criminally liable. In evaluating whether there is a deviation from responsible conduct that imposes criminal liability, the courts assess whether public law requirements are complied and whether the acts are in accordance with industry practice.[6]

---

[2] Ref. strl. § 281.
[3] Look at LB-2009-55987.
[4] Ref. straffeloven § 280.
[5] Ref. aksjeloven §§ 6-12 and 6-13. Look also at Rt. 2013 p. 312 premiss 27 flg.
[6] Example: Rt. 2013 s. 312.

Acknowledgment

The authors would like to thank especially Roger Flage at University of Stavanger (Norway) for input and comments to an earlier version of the present paper, Pöyry, Dala Vind, Fortum, Satavia, Norkring, TechnoCentre éolien (Wind Energy TechnoCentre), DNV-GL, Jenny Lunden, Bengt Göransson, Marc Noël de Wild, Matthew Homola, Erik Aretorn, Niklas Lundaahl, Gaute Gruben, Göran Ronsten, Adam Durant, Frank Børre Pedersen, Sonia Liléo, Arne Olsen, and other colleagues for discussions and contributions to the presented work.

The Norwegian contribution to this work has been supported by The Norwegian Water Resource and Energy Directorate and through the project "Tackling icing issues for wind energy production", project number 269436/E20, which is partly funded by the Research Council of Norway through the ENERGIX program. In addition the authors would like to thank Statkraft for sponsoring the Norwegian participation in IEA Wind Task 19 for the working period 2016-2018. The monitoring campaign used to generate specimens was a project funded by the Austrian Climate and Energy Fund (KLI.EN). Further thanks go to Energie Burgenland Windkraft who made their turbines available for drop experiments.

References

[1] Krenn et al. 2017. IEA Wind RD&D Task 19 – Wind Energy in Cold Climates. *Guidelines for Ice-fall / ice-throw risk assessments.* https://www.ieawind.org/task_19/task_19_riskAssessment.html

[2] Krenn et al. (2017) IEA Wind Task 19*: Standardisation of input parameters for ice throw / ice fall risk assessments*. WindEurope Conference 2017 (Amsterdam).

[3] Weber, N., Barup, S., Rodriguez, H., Cornmeau, M., Zwölfer-Dorau. U. (2017). Making ice fall and ice throw predictions for wind turbines more reliable. WindEurope Conference, Amsterdam 2017. Also: Wind Turbine Icing and Ice Prevention Forum. Bis group. Berlin Germany 6-7th November 2017.

[4] Bredesen, R.E., Bourgeois, S., Haaheim, D. (2017). IEA Wind RD&D Task 19. *Open workshop for the exchange of experience and practices regarding ice throw and ice fall for the upcoming guidelines.* Winterwind, February 15, 2017

[5] Bredesen, R.E. (2017*). Icethrow from wind turbines - Assessment and Risk Management.* Winterwind conference 2017. http://winterwind.se/wp-content/uploads/2015/08/3_3_28_Bredesen_IEA_Task_19_-_IceRisk_Review_of_current_knowledge_and_the_way_forward_in_risk_assessments_associated_with_ice_throw_from_wind_turbine_blades_Pub_v1-1.pdf

[6] Burgeois S., (2017). *IEA Wind Task 19 - Workshop on Ice throw, Winterwind 2017*, Skellefteå, Sweden, February 2017, http://windren.se/WW2017/3_3_28_Bourgeois_IEA_Task_19_-_IceRisk_Review_of_current_knowledge_and_the_way_forward_in_risk_assessments_associated_with_ice_throw_from_wind_turbine_blades_Pub_v1.pdf

[7] Bredesen, R.E., Flage, R., Butt, B. (2018). *Norwegian guidelines regarding the risk of icethrow for the public*. Winterwind 2018. Accepted.

[8] Flage, R. (2017). *Note to NVE regarding risk related to ice throw and ice fall from wind turbines.*

[9] NVE (2018). *Tiltak for å unngå skader fra iskast og isnedfall fra vindturbiner - Veilder til gjennomføring av risikovurdering og implementering av tiltak for å håndtere risiko for skader fra iskast og isnedfall i norske vindkraftverk.* The Norwegian Water Resource and Energy Directorate (NVE).

[10] Ronsten. G. (2017). Wind Energy icing research and activities in Sweden. WWEC 2017. World Wind Energy Conf. Malmö, Sweden. https://wwec2017.com/wp-content/uploads/2017/06/WWEC2017_Full_Paper_Ronsten_A4_v11.pdf

[11] ISO 12494:2017 (en). Atmospheric icing of structures - Second edition. The International Organization for Standardization. https://www.iso.org/obp/ui/#iso:std:iso:12494:ed-2:v1:en

[12] Wadham-Gagnon, M. (2013). *Ice profile Classification Based on ISO 12494. Winterwind 2013*

[13] IEA Wind. 2017. *Expert Group Study on Recommended Practices. 13. Wind Energy in Cold Climates. 2. Edition.* https://www.ieawind.org/task_19/020617/Task%2019%20RP%2013_v2%20report_2017_ExCo%20appr%201Feb17_FINAL.pdf

[14] Seifert, H.; Westerhellweg, A. & Krning, J. (2003), 'Risk analysis of ice throw from wind turbines', BOREAS 6(9).

[15] Energimyndigheten. (2017). *ICETHROWER – mapping and tool for risk analysis.* Slutrapport. http://www.energimyndigheten.se/forskning-och-innovation/projektdatabas/sokresultat/?projectid=18386

[16] Pöyry (2017), 'ICETHROWER Database and Software', online Database, Available online https://onepoyry.sharepoint.com/sites/8H50156.100/_layouts/15/guestaccess.aspx?folderid=0f4bf8e9ab7ea4ce88996e2adb4b602e8&authkey=AasdYcr1qoxUoXaO96rvrjo&expiration=2018-08-28T06%3a59%3a08.000Z

[17] Lunden, J. (2017), 'ICETHROWER - Mapping and tool for risk analysis', presentation at Winterwind 2017.

[18] Göransson, B. Lundén, J., Hultin, K., Aretorn, E., Sundström, J., Odemark, Y., Montgomerie, B., (2017). ICETHROWER - ICE THROWER Evaluation and Risk Analysis Tools. Pöyry Sweden. http://www.energimyndigheten.se/forskning-och-innovation/projektdatabas/sokresultat/?projectid=18386

[19] Drapalik, M. (2017), 'Risiken der Vereisungproblematik bei Windenergieanlagen unter besonderer Berücksichtigung der Situation in Österreich', PhD thesis, Universität für Bodenkultur Wien.

[20] Drapalik, M. (2017), 'Risk of ice throw from SWT', presentation at the 3. Internationale Kleinwindkrafttagung 2017, Vienna.

[21] Drapalik (2017) Experimental investigation of risk from ice throw and ice shed. Winterwind 2017. http://winterwind.se/wp-content/uploads/2015/08/3_1_03_Drapalik_Experimental_investigation_of_risk_from_ice_throw_and_ice_shed_Pub_v1.pdf

[22] Drapalik, M., Bredesen, R.E. (2017), Experimental investigation of risk from ice shed and ice throw. World Wind Energy Conference Malmö, Sweden. WWEC2017. https://wwec2017.com/wp-content/uploads/2017/06/Drapalik-Bredesen-Experimental-validation-of-models-for-ice-shed-risk-analysis.pdf https://wwec2017.com/wp-content/uploads/2017/06/Drapalik-Experimental-validation-of-models-for-ice-shed-risk-analysis-v6.pptx

[23] Bredesen, R.E., Refsum, H.A., 2015. Methods for evaluating risk caused by ice throw and ice fall from wind turbines and other tall structures. Kjeller Vindteknikk, NO (38). IWAIS 2015. 16th International workshop on atmospheric icing of structures. ISBN 978-91-637-552-8. Uppsala, Sweden, June 28 - July 3 2015. www.iwais.org. http://www.vindteknikk.no/_extension/media/271/orig/38_13_03_Paper_Bredesen_Methods_for_evaluating_risk_caused_by_ice_throw_.pdf

[24] Bredesen (2017) IceRisk: Probability of ice throw from wind turbines - Description of methodology with validation using the IceThrower database. KVT Memorandum KVT/REB/2017/N033

[25] Bredesen, R.E. (2017). Amsterdam, The Netherlands. IceRisk: Probability of ice throw from wind turbine blades. Report number: KVT/REB/2017/R073. Kjeller Vindteknikk report.

[26] Biswas, S.; Taylor, P. & Salmon, J. (2012), A model of ice throw trajectories from wind turbines', Wind Energy 15(7), 889--901.

[27] Harstveit, K. 1996. Full scale measurements of gust factors and turbulence intensity, and their relations in hilly terrain. *Journal of Wind Engineering and Industrial Aerodynamics 61,* 1996, pp 195-205.

[28] Amundrud, Ø. & Aven, T. (2015), 'On how to understand and acknowledge risk', *Reliability Engineering and System Safety* **142**(Complete), 42-47.

[29] Aven, T. & Renn, O. (2010), *Risk Management and Governance*, Springer Berlin Heidelberg.

[30] Renn, O. in Heath, R. L. & O'Hair, H. D., ed., (2010), *Handbook of Risk and Crisis Communication*, Routledge, chapter 4: Risk Communication: Insights and Requirements for Designing Successful Communication Programs on Health and Environment Hazards, pp. 80-98.

[31] Seifert, H.; Westerhellweg, A. & Kröning, J. (2003), 'Risk analysis of ice throw from wind turbines', *BOREAS* **6**(9).

[32] Biswas, S.; Taylor, P. & Salmon, J. (2012), 'A model of ice throw trajectories from wind turbines', *Wind Energy* **15**(7), 889--901.

[33] DIN EN 50126 (2016), 'DIN EN 50126-1:2016-03; VDE 0115-103-1:2016-03 - Bahnanwendungen - Spezifikation und Nachweis von Zuverlässigkeit, Verfügbarkeit, InstanSummary

[34] Hahm, T. og Stoffels, N. (2016). *Ice Throw Hazard. Experiences and Recent Developments in Germany.* http://f2e.de/files/public/f2e_ice_throw_hazard_windeurope2016.pdf

[35] Nöel de Wild, M. (2017). Master thesis, Uppsala University Publications: A view into future potential ice throw policies and their effect on the yield of a virtual wind farm. http://uu.diva-portal.org/smash/record.jsf?pid=diva2%3A1134442&dswid=7435

[36] DNV GL (2016). Enabling confidence: addressing uncertainty in risk assessments. DNV GL Strategic Research & Innovation position paper. Tilgjengelig fra: https://www.dnvgl.com/oilgas/download/position-paper-enabling-confidence-addressing-uncertainty-in-risk-assessment.html

[37] Cattin, R. (2012). *Ice trow studies. Glütsch and St. Brais.* https://www.slideshare.net/WinterwindConference/ice-throw-reloaded-studies-at-guetsch-and-st-brais

[38] Norconsult AS. (2016). *Samfunnsmessige virkninger av vindkraftverk. En etterprøving av fire vindkraftverk.* https://www.norconsult.no/contentassets/ebf62d613b214d1cb9b4c82d342952d1/samfunnsmessige-virkninger-av-vindkraft---en-etterproving-av-fire-vindkraftverk.pdf

[39] Tammelin, B., Cavalieere, M., Holttinen, H., Morgan, C., Seifert, H. og Säntti, K. 1996-1998. WIND ENERGY PRODUCTION IN COLD CLIMATE (WECO). http://cordis.europa.eu/documents/documentlibrary/47698271EN6.pdf

[40] Sarlak, H., Sørensen, J.N. (2016) Analysis of throw distances of detached objects from horizontal-axis wind turbines. Wind Energ. 2016; 19:151–166

[41] Xue, H., Khawaja, H. (2016). Review of the Phenomenon of Ice Shedding from Wind Turbine Blades. Int Jnl of Multiphysics Volume 10. Number 3. 2016.

[42] Amundrud, Ø., & Aven, T. (2015). On how to understand and acknowledge risk. *Reliability Engineering & System Safety*, 142, 42-47.

[43] Aven, T. (2013). Practical implications of the new risk perspectives. *Reliability Engineering & System Safety*, 115, 136-145.

[44] Aven, T. (2015). *Risikostyring*. 2. utg. Universitetsforlaget.

[45] Aven, T., & Renn, O. (2010). Risk Management and Governance. Concepts, Guidelines and Applications. Berlin: Springer.

[46] Aven, T., & Vinnem, J. E. (2005). On the use of risk acceptance criteria in the offshore oil and gas industry. *Reliability Engineering & System Safety*, 90(1), 15-24.

[47] Berner, C., & Flage, R. (2016). Strengthening quantitative risk assessments by systematic treatment of uncertain assumptions. *Reliability Engineering & System Safety*, 151, 46-59.

[48] Bier, V. M. (2001). On the state of the art: risk communication to the public. *Reliability Engineering & System Safety*, 71(2), 139-150.

[49] Bjerga T, Aven T & Flage R. (2017) Completeness uncertainty: Conceptual clarification and treatment. In T. Aven & E. Zio (Eds.), *Knowledge in Risk Assessments*. Wiley.

[50] Flage, R., & Aven, T. (2009). Expressing and communicating uncertainty in relation to

quantitative risk analysis. *Reliability & Risk Analysis: Theory & Application*, 2(13), 9-18.
[51] Funtowicz, S. O., & Ravetz, J. R. (1990). *Uncertainty and Quality in Science for Policy*. Springer.
[52] ISO 31000 (2009). Risk Management – Principles and Guidelines.
[53] NORSOK Z-013 (2010). Risk and Emergency Preparedness Assessment. Standard Norge.
[54] Petroleumstilsynet (u.å.) Risiko og risikoforståelse. Tilgjengelig fra: http://www.ptil.no/risiko-og-risikoforstaelse/risiko-og-risikoforstaelse-article9600-823.html.
[55] Renn, O. (2008). Risk Governance: Coping with Uncertainty in a Complex World. Earthscan.
[56] SRA (2015). SRA glossary. Society for Risk Analysis. Tilgjengelig fra: http://www.sra.org/sites/default/files/pdf/SRA-glossary-approved22june2015-x.pdf.
[57] Aven (2015) Risk analysis. Wiley. 2nd ed.
[58] Aven, T. (2014) Risk, Surprises and Black Swans: Fundamental Ideas and Concepts in Risk Assessment and Risk Management. London: Routledge.
[59] Flage, R. and Aven, T. (2009) Expressing and communicating uncertainty in relation to quantitative risk analysis (QRA). Reliability and Risk Analysis: Theory and Applications, 2(13), 9-18.
[60] Terje Aven - Risikostyring 2. utgave (2015)
[61] Terje Aven, Willy Røed, Hermanns. Wiencke - Risikoanalyse 2. utgave (2017)
[62] Aven, T. (2009) Risk analysis and Management
[63] Baker, C. J. (2007), 'The debris flight equations', Journal of Wind Engineering and Industrial Aerodynamics 95(5), 329--353.
[64] Center for Chemical Process Safety) CCPS, (C. (2009), Guidelines for Developing Quantitative Safety Risk Criteria, Wiley-Blackwell.
[65] DSB Temaveileder "Sikkerheten rundt anlegg som håndterer brannfarlige, reaksjonsfarlige, trykksatte og eksplosjonsfarlige stoffer: Kriterier for akseptabel risiko", Mai 2013.
[66] TNO Greenbook, Methods for the determination of possible damage, to people and objects resulting from release of hazardous materials, CPR 16E, TNO – The Netherlands Organisation of Applied Scientific Research, first edition 1992.
[67] Range Commanders Council (RCC). Standard 321-07. Common risk criteria for national test ranges: Inert debris.
[68] European Aviation Safety Agency. Introduction of a regulatory framework for the operation of drones https://www.easa.europa.eu/system/files/dfu/NPA%202017-05%20%28A%29_0.pdf
[69] la Cour-Harbo, A. (2017). Ground impact probability distribution for small unmanned aircraft in ballistic descent. Preprint submitted to Reliability Engineering & System Safety June 18, 2017.
[70] la Cour-Harbo, A. (2017). Quantifying risk of ground impact fatalities of power line inspection BVLOS flight with small unmanned aircraft. 10.1109/ICUAS.2017.7991323.
[71] Homola, M.C. (2011). *Atmospheric icing on wind turbines: Modeling and consequences for energy production* (Doctoral dissertation). Retrieved from https://brage.bibsys.no/xmlui/handle/11250/260519

IOP Conf. Series: Journal of Physics: Conf. Series **926** (2017) 012002 doi:10.1088/1742-6596/926/1/012002

Ice-Release and Erosion Resistant Materials for Wind Turbines

Wei Zhang*, Cameron Brinn, Alex Cook, Fernando Pascual-Marquez

Polymer Exploration Group LLC, Ashland VA 23005, USA.

*wzhang@pegllc.org

Abstract. Icing conditions may cause wind turbine generators to partially lose productivity or to be completely shut down to avoid structural damage. At present, commercially available technologies to mitigate this problem consist of expensive, energy hungry heating elements, which costs roughly 70,000 euro per medium size turbine. Conventional passive ice protection coating systems heavily rely on delicate surface structures and expensive materials to create water repellent superhydrophobic / low surface energy surfaces, which have been proven to be ineffective against ice accumulation. The lack of performance among conventional ice protection materials stems from a flaw in the approach to the problem: failure to recognize that water in its liquid form (WATER) and water in its solid form (ICE) are two different things. Something that works for WATER does not automatically work for ICE. Another reason is that many superhydrophobic materials are often reliant upon often fragile micro-structured surfaces to achieve their intended effects. This paper discusses a fundamentally different approach to the creation of a robust, low cost, durable, and multifunctional materials for ice release and erosion resistance. This National Science Foundation sponsored ice-release coating technology holds promise for protecting wind turbine blades and towers, thus potentially increasing reliability for power generation under icing conditions. Because of the vulnerability of wind turbine blades to ice buildup and erosion damages, wind farm facilities stand to reap considerable benefits.

1. Introduction: An Approach different from conventional wisdom

Conventional wisdom suggests that if water does not stick to the surface, ice "should not". Therefore, there has been much research on the development of the superhydrophobic surfaces, which have been proven ineffective against ice accumulation. Many superhydrophobic materials are reliant upon micro-structured surfaces (mimicking lotus leaves[3]) to achieve their intended effects: water beads up and rolls off the surface. Not only are these surfaces expensive to fabricate and are typically fragile, but they are counterproductive in many icing conditions, e.g. frost.[4, 5] For the example of frost, ice particles may form in between the microsctructures and the microsstructured surfaces would reinforce the adhesion of ice on the surface instead of reducing ice adhesion. Some of these superhydrophobic surfaces are reported with very good ice protection, especially in laboratory testing. Water used for these studies are all above freezing point because of considerations for handling in the laboratory. Under these laboratory testing conditions, water droplets contact surfaces in liquid form, the surface tension allows it to form spherical shapes before freezing. This process, may only take miliseconds, is enough for it reduces the actual contact area of the water/ice with surface, as illustrated in Figure 1 (left). As a result, low apparent ice adhesion forces may be recorded and it may be perceived as good ice protection. This can be considered as front end ice protection: release of water droplets before ice is formed, or reducing contact area to reduce ice adhesion force on the surface.

IOP Conf. Series: Journal of Physics: Conf. Series **926** (2017) 012002 doi:10.1088/1742-6596/926/1/012002

In real world applications, such as gusting wind carrying super-cooled water droplets that impact a cold surface, the performance can be dramatically different. Such conditions do not allow for the water droplet to stay in liquid form long enough to form beads and to reduce the surface contact area, as illustrated in Figure 1 (below). The impacting water droplets instantly form ice upon contact with the surface. After ice has been formed, the ice protection performance is defined strictly by the adhesion strength of the ice. The several contributing factors and various means to reduce the ice adhesion strength are discussed below. Because the surface energy is only one of the several key parameters, it was discovered that many hydrophobic and superhydrophobic coatings, including many fluoropolymers, do not offer adequate protection against ice under real world icing conditions.[6, 7]

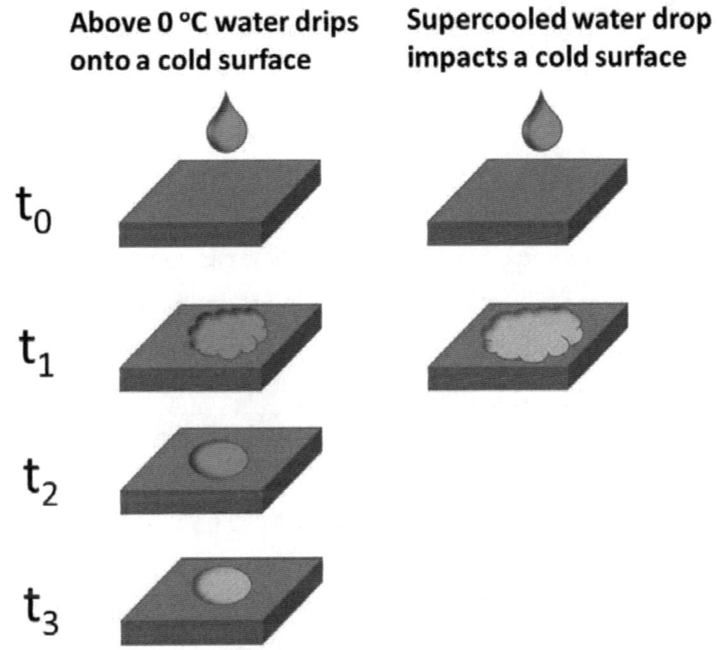

Figure 1. Illustration of ice formation on hydrophobic surfaces.

We have examined the situation from multiple angles; we recognize the fundamental differences between ice and water, account for the contributions of both surface science, and the mechanics of the ice removal process. Based on the findings from this National Science Foundation supported research, we developed a new ice release model that better describes the icing phenomenon.

This model, as illustrated in Figure 2 and the Kendal Equation (Eq 1), accounts for the contributions of multiple factors to the overall ice release performance of a surface (not just surface energy) because the ice release process is not a purely a surface phenomenon.[1, 2, 8-14] Of these four major factors: (1) work of adhesion (w_a), (2) modulus (K), and (3) functional thickness (t), and (4) surface area (A), the first three are intrinsic to the surface. Reduced actual contact area of ice with a surface is why some hydrophobic surfaces may display good ice protection under certain controlled icing conditions, shown in Figure 1A and discussed above.

Work of adhesion (w_a), directly relates to the surface tension or surface energy (or interfacial energy between ice and the surface) is the main focus of most of the research and development work done currently. Consider perfluorohexane which has a surface tension of 11.9 mN/m,[15] it largely set the lower limit for the surface tension of any solid materials, let alone the materials that can be used on a wind turbine blade. Typically, wind turbine blades are made of composites, so that the upper limit of the surface tension range is around 45 mN/m,[15] that of epoxy (which is considered one of the polymeric materials with the highest surface tension). This means the maximum difference that the work of adhesion can make to the ice adhesion force is a reduction by a factor of 2 (square root of 45/11.9).

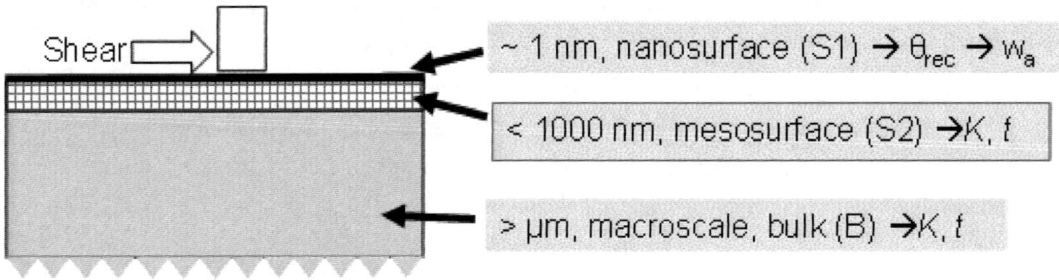

Figure 2. The release mechanism of rigid adherent on elastomeric coating. Low modulus mesosurface contributes to the peak removal force P_s with two of the three parameters in Kendal's theory (Eq 1): (1) work of adhesion (w_a), (2) modulus (K), and (3) thickness (t). [1, 2]

Kendall theory adapted to ice removal: Peak Removal Force in shear is designated P_s

$$P_s \propto A\left(\frac{2w_aK}{t}\right)^{1/2} \quad \textbf{Eq 1}$$

P_s: Peak removal force

Modulus (K) of composite materials varies from a few GPa to tens of GPa depending on the fiber type and content and those of the matrix material. If such a surface is coated with an elastic rubbery materials, the modulus at the surface can be reduced to a few MPa. By varying the modulus of the surface material, the ice adhesion strength can be reduced by 10 to 100 times (square root of 100-10000). It should be noted that the above two parameters cannot be tested independently because it is impossible to create materials by changing only one of these two parameters.

Functional thickness (t) of a surface material is the other factor that can be manipulated in order to reduce the ice adhesion, although the authors believe t is not simply the coating/material thickness and it is not independent from K. In addition, when t is high, the additional load (mass) added to a wind turbine has to be considered.

Based on the above model and the consideration of multiple factors, we have developed a new passive ice release coating system, Easy-Ice-Release-Coating (EIRC), and it has been proven effective with laboratory tests. With the EIRC demonstrates low ice adhesion strength and good bonding strength to a variety of substrates including the commonly used fiber-reinforced composites by wind turbine blade manufacturers. In addition, these ice-release materials can be used as leading edge protection coating to protect WTG blades against sand and rain erosion.

2. Material preparation

2.1. Materials

Materials for creating the ice release coatings and materials are all commercially available engineering polymeric materials.[8] Poly(tetramethylene) oxide 2000 g/mol (PTMO-2K) and dibutyltin dilaurate (T-12) were purchased from Aldrich. Isocyanatopropyltriethoxysilane (ICP-Si(OEt)3), hydroxyl terminated poly (polydimethylsilane) (Silanol), and bis(triethoxysilyl)-ethane (BTESE) were purchased from Gelest, Inc. Tetrahydrofuran (THF) was purchased from Acros Chemicals. Thermoplastic polyurethane, Estane ALR E72A was purchased from Lubrizol. Isopropanol (IPA) was purchased from Fisher Scientific. 1-Butanol was purchased from Green Biologics Inc. Materials were used as received except the thermoplastic polyurethane resin, which was purified prior to use for the ice release coatings. The purification process removes unreacted monomers and other small molecules that would interfere with the ice release coating.

2.2. Coating Preparation
The coating preparation process followed a four-step process: 1. preparation of hybrid formation precursors by reacting the respective components at elevated temperatures. 2. preparation of the coating solution by mixing the polyurethane stock solution, 3. coating application by drip coating (which can easily converted to brush or spray coating), and 4. drying and curing.[8] It should be noted that the application, drying and curing parameters are very important as they determine whether a sufficient degree of self-stratification has occurred, and whether the unique mesosurface structure has fully formed. The drying and curing conditions will be disclosed at a later date.

3. Pilot Production Process
A pilot processing equipment was developed to 1) prepare larger quantities of materials for field tests, 2) to check the feasibility of scaling up, and 3) to collect parameters and processing conditions for full industrial scale production. Ice release materials were created by casting on the substrates. The casted film on the substrates is then fed into a chamber and remains in the chamber for a certain time. The drying and curing conditions of the chamber is controlled so to match the batch processing conditions described in the previous section. The dried film is then collected onto another roll before testing. The materials prepared with this pilot production process was tested for their ice release performances using one of the testing methods discussed below. The results are also compared to the laboratory coating preparation methods.

4. Ice Release Test
In order to test the adhesion strength of ice on various surfaces, three high throughput methods were developed in our labs to measure the shear stress necessary to remove icing from a given surface. Slightly modified commercial analytical instrument or in-house developed instrument are used for these methods. These methods, while unique, share some fundamental similarities. In the absence of standardized methods for testing ice adhesion, there are a number of different custom-made testing devices reported.[16-21] We have taken inspiration from these devices and those used by Swain[22] for different, but related purposes, and created a proprietary testing method. These tests are done by a probe that is parallel to the coating surface to apply a shear force to remove the ice adherents on the surface. These methods have been described in a recent publication.[9-14, 23]

Each method makes use of a sample surface (the ice release performance of which is being measured), and an ice form (an open ended prism with a known cross sectional area). The sample is cooled to the desired testing temperature and the ice form is placed upon it with the open cross section facing down. The ice form at the interface of the ice holder and sample. A force is applied to the ice via the probe in a direction parallel to the ice-sample interface. The area of the interface, and the force required to remove the ice sample are recorded and used to calculate the shear stress (reported as peak removal force below) at interfacial failure. Cohesive failure (ice cylinder breaks rather than detaches from surface) was occasionally observed, especially for control (uncoated) samples. But the accurate and delicate instrumentation employed to carry out the ice release test usually reached the force limit of the machine before causing cohesive failures and some adhesive failures. It should be pointed out that a surface is not useful as ice release coating if cohesive failure is observed. However, because of the unknown values for the control samples where instrumentation maxed out before causing interfacial/cohesive failures, the relative increase in ice release performance (ratio of the peak removal forces of the control surface versus that of the sample) may only be an estimated value.

The materials were then tested for ice release performance using two independent and third-party validated *in-situ* ice adhesion test instruments. These two instruments are featured with high accuracy but low throughput and high throughput but low accuracy. However, over a large data set, the averages agree with each other very well while the main difference is the error bar. The in-situ testing methods using the same setup for ice formation and testing without transfer of iced specimens from one place to another. It eliminates the possibility of thermal or mechanical shock during sample transfer (between the freezing chamber and the testing chamber) affecting the ice adhesion force. Additionally, these methods ensure that the forces applied to the sample are true shear forces to minimize the effects of forces other than shear acting on the ice.

Third party testing was performed at the Adverse Environment Rotor Test Stand (AERTS) facility in the Aerospace Engineering Department at Penn State University. The rotor beam is shown with coated coupons mounted on the end of test beam leading edge. Rotation is initiated up to 1000 rpm after the room is cooled to the desired temperature. The tip speeds are up to ~470 ft/s or 320 MPH. After reaching the desired tip speed, a cold water mist is sprayed into the room. Water droplets freeze on the surface after impinging on the rotating sample and ice accumulates on the surface. As the mass grows, the shear force at the interface of ice and test surface increases until the ice flies off. Measurements obtained from a strain gauge and torque sensor are used to calculate the numerical shear removal force. This test was designed to reproduce natural icing conditions such as those on a helicopter rotor or airplane wing. However, because of the relatively low rotor speed the test better simulates ice release from wind turbine blades, which is a near-term targeted application. Of course, the Penn State test can stand alone as a valid general ice-release testing method without a specific targeted application.[16, 17, 24]

5. Discussion

Ice adhesion strength on the surface is shown to greatly reduce ice adhesion of ~ 1 MPa on multiple substrates (steel, aluminum, composites etc.) down to ~50 kPa at -10 °C. Good adhesion was observed on these substrates as well. The results were achieved by the formation of a self-segregated surface structure with nanosurface, mesosurface and bulk coating. This is demonstrated through experimental manipulation of the curing conditions of the EIRC. When conditions are such that the mesosurface is allowed to form, the ice adhesion is very low. When cure conditions are altered such that the mesosurface is precluded from forming, the ice adhesion is much more similar to that of the unmodified polyurethane bulk component of the EIRC.

The EIRC prepared in the pilot production process aligns very well with the other preparation methods, which means the mass production can be implemented by scaling up the pilot production process. A more detailed ice removal data chart is shown in Figure 4 below. Except a couple outliers, most of the ice cylinders were removed at a peak force of less than 60 kPa and the overall average is ~50 kPa.

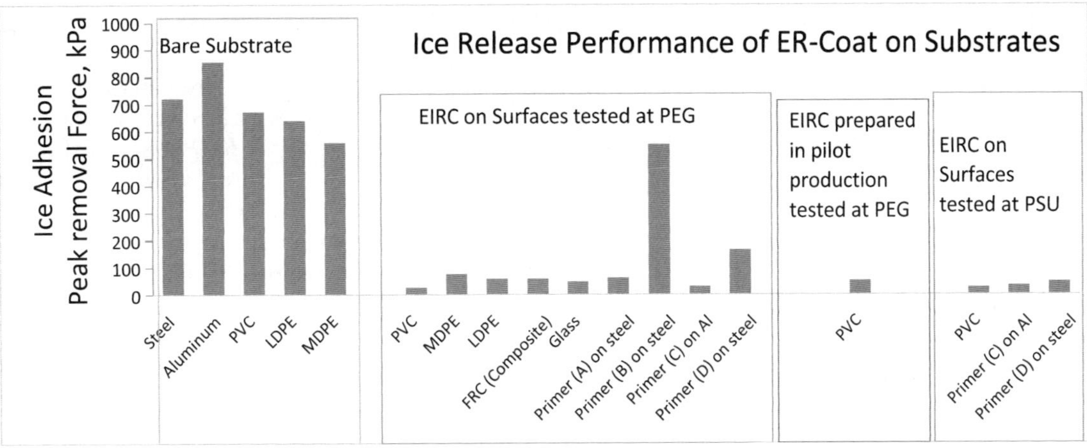

Figure 3. Ice release performance of substrates coated with EIRC.[2] The results in the left box are bare substrate surfaces (most of these tests did not register a release from surface – interfacial failure, instead ice cylinder broke – cohesive failure), the middle box is for ice release tests done at PEG, and the right box is for tests carried at Pennsylvania State University.[16, 17, 24]

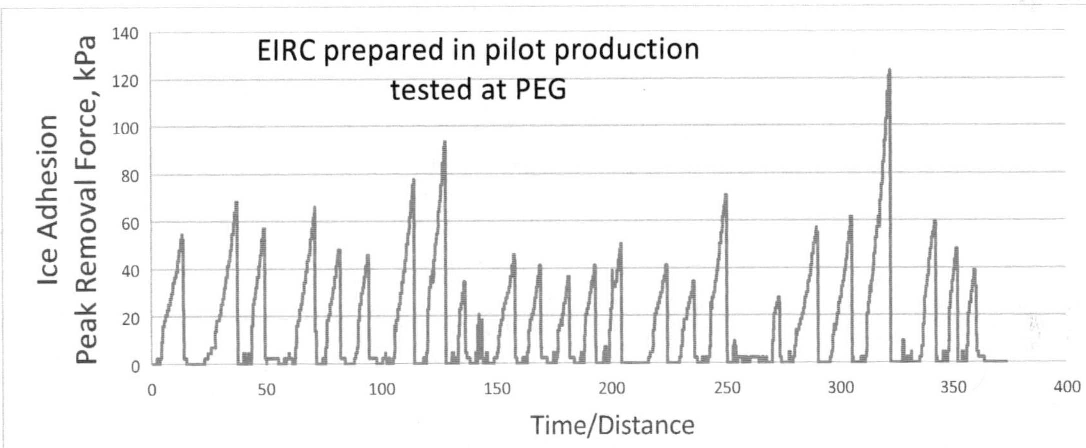

Figure 4. Ice release test using one of the PEG *in-situ* ice removal test equipment. Multiple ice cylinders were formed on the surface and then removed using a force probe. The force, the time/distance is recorded during the removal process. Because multiple ice cylinders were removed in a single test, the x-axis is used to differentiate the different ice cylinders and the actual number is meaningless in this case.

6. Rain Erosion Test

The kinetic energy in an impacting sand particle or a rain droplet has to be dissipated or absorbed one way or another. If not absorbed and dissipated in a nondestructive manner, the particles with high energy will cause cracks, wear, abrasion or other types of mechanical damages (erosion). Given time, the airfoil of the surface can be greatly damaged due to the erosion. A material with high strength, high elasticity, but low modulus has a good potential for good erosion resistance.[25] The requirements for good erosion resistance properties happened to agree in a great deal with that of the ice release materials we have developed. Therefore, the ice release materials were also tested for rain erosion performance. At the same time, an ice release material without good erosion resistance would have a longevity issue, especially in the wind energy and aerospace industries where erosion is a big problem.

Based on the ASTM G73-10, Standard Test Method for Liquid Impingement Erosion Using Rotating Apparatus, a modified and simplified rain erosion test was developed in our laboratory to preliminarily test the erosion resistance. This test was not designed to replace the accredited third party test as the velocity of the water droplets are not calibrated. However, the ice release materials were tested side-by-side with the benchmark of erosion protection tape on the market (3M W8607 Wind Protection Tape).

The results are shown in Figure 5 from the same erosion conditions. From this side-by-side comparison, EIRC fared very well compared with this wind protection tape. At the same time length, EIRC coating was damaged and the substrate was exposed, while the W8607 was damaged. From this preliminary test, the difference cannot be differentiated. It was discovered that defects and air pockets formed during the coating application process may cause an early failure in the coating.

Third party rain erosion testing are underway. Results will be reported at a later time when they become available.

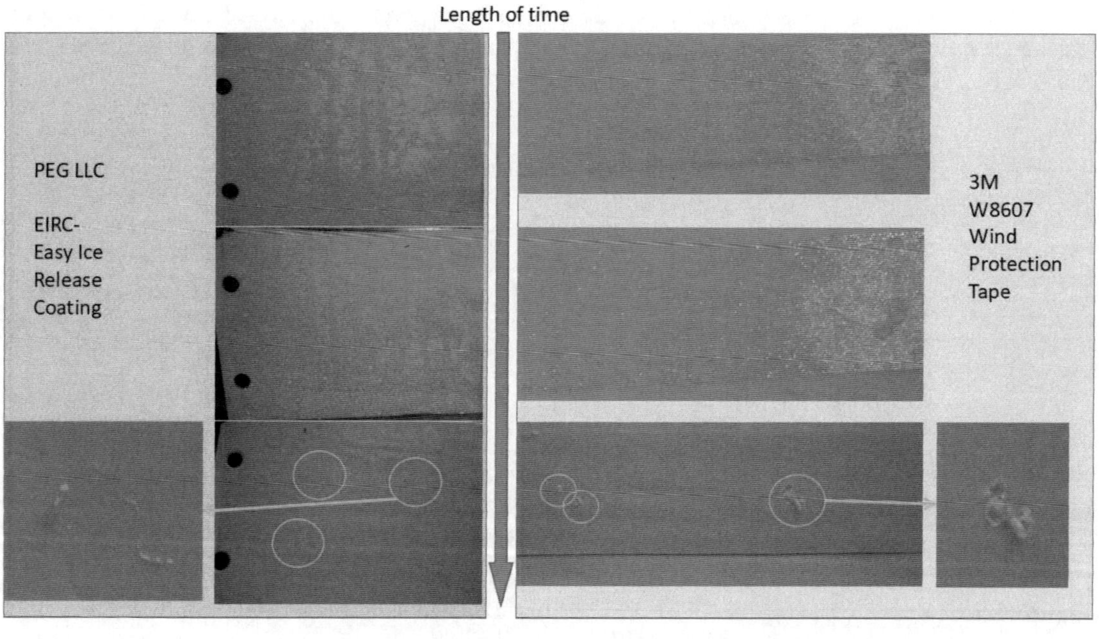

Figure 5. Preliminary rain erosion test side-by-side with a benchmark product

7. Conclusion

The problem of ice adhesion, in its simplest form, can be reduced to the mechanical interaction between two substances: the ice itself, and the surface upon which it is adherent. The mechanical properties of each substance will inform their behavior during the mechanical process of ice removal. Conventionally, superhydrophobic coatings have been promoted for the purpose of creating "ice-phobic" surfaces. This ignores the mechanical contribution to the ice release process and focuses only on the surface energy. As has been shown, low surface energy alone is insufficient to reduce ice adhesion to a meaningful degree. The new EIRC technology takes into account multiple contributing factors to ice adhesion. The resulting coating system not only provides ice protection, but also protects wind turbines from erosion damage. The coating does not trade off rain erosion protection for ice protection, or vice versa. The consideration of the mechanical components of an ice release process lead us to the development of a mechanically strong material with a low modulus and high elasticity.

The mechanical properties of the coating coincidentally meet the requirements for good erosion resistance. Also because of the toughness of the coating, the ice release properties are not easily compromised by erosion of coating surface, as is the case for many other ice protection materials.

References

[1] K. Kendall. "The adhesion and surface energy of elastic solids" *Journal of Physics D: Applied Physics* 4 (1971): 1186.
[2] C. Brinn, K. Brunson, D. Klein, K. J. Wynne, W. Zhang. "Shaking off the cold: Ice-Release Coatings" *European Coatings Journal* 2017 (2017): 52.
[3] Z. Q. Yuan, J. P. Bin, X. Wang, Q. L. Liu, D. J. Zhao, H. Chen, H. Y. Jiang. "Preparation and anti-icing property of a lotus-leaf-like superhydrophobic low-density polyethylene coating with low sliding angle" *Polymer Engineering and Science* 52 (2012): 2310.
[4] K. K. Varanasi, T. Deng, J. D. Smith, M. Hsu, N. Bhate. "Frost formation and ice adhesion on superhydrophobic surfaces" *Applied Physics Letters* 97 (2010).
[5] S. Jung, M. Dorrestijn, D. Raps, A. Das, C. M. Megaridis, D. Poulikakos. "Are Superhydrophobic Surfaces Best for Icephobicity?" *Langmuir* 27 (2011): 3059.
[6] C. C. Ryerson, (Ed: C. R. R. a. E. L. Us Army Corps of Engineers, U.S. Army Engineer Research and Development Center), April 2009 2009.
[7] U. A. C. o. Engineers, (Ed: D. o. t. A. Us Army Corps of Engineers), 30 September 2006 2006.
[8] K. J. Wynne, W. Zhang, *United States Patent 20160215167*, 2016.
[9] C. Wang, W. Zhang, A. Siva, D. Tiea, K. J. Wynne. "Laboratory Test for Ice Adhesion Strength Using Commercial Instrumentation" *Langmuir* 30 (2014): 540.
[10] C. Wang, T. Fuller, W. Zhang, K. J. Wynne. "Thickness Dependence of Ice Removal Stress for a Polydimethylsiloxane Nanocomposite: Sylgard 184" *Langmuir* 30 (2014): 12819.
[11] K. J. Wynne, W. Zhang, C. Wang, S. Chakrabarty, *WO2013016594A2*, 2013.
[12] K. J. Wynne, S. Chakrabarty, W. Zhang, C. Wang, "Hybrid ice-release coatings: New polyoxetane fluorous elastomers, new laboratory test, new theory", 2013.
[13] S. Chakrabarty, C. Wang, W. Zhang, K. J. Wynne. "Rigid Adherent-Resistant Elastomers (RARE): Nano-, Meso-, and Microscale Tuning of Hybrid Fluorous Polyoxetane–Polyurethane Blend Coatings" *Macromolecules* 46 (2013): 2984.
[14] W. Zhang, S. Chakrabarty, K. J. Wynne, "Ice-phobic coatings: New theory, new laboratory test, and new hybrid fluorous polyoxetane systems", 2012.
[15] http://www.surface-tension.de/, retrieved September 2015.
[16] A. M. Palacios, J. L. Palacios, L. Sanchez. "Eliciting a human understandable model of ice adhesion strength for rotor blade leading edge materials from uncertain experimental data" *Expert Systems with Applications* 39 (2012): 10212.
[17] J. L. Palacios, Y. Han, E. W. Brouwers, E. C. Smith. "Icing Environment Rotor Test Stand Liquid Water Content Measurement Procedures and Ice Shape Correlation" *J. Am. Helicopter Soc.* 57 (2012): 12.
[18] V. K. Croutch, R. A. Hartley. "Adhesion of ice to coatings and the performance of ice release coatings" *J. Coat. Technol. FIELD Full Journal Title:Journal of Coatings Technology* 64 (1992): 41.

[19] A. J. Meuler, J. D. Smith, K. K. Varanasi, J. M. Mabry, G. H. McKiney, R. E. Cohen. "Relationships between Water Wettability and Ice Adhesion" *Acs Applied Materials & Interfaces* 2 (2010): 3100.
[20] H. H. G. Jellinek, H. Kachi, S. Kittaka, M. Lee, R. Yokota. "Ice releasing block-copolymer coatings" *Colloid and Polymer Science* 256 (1978): 544.
[21] S. Farhadi, M. Farzaneh, S. A. Kulinich. "Anti-icing performance of superhydrophobic surfaces" *Applied Surface Science* 257 (2011): 6264.
[22] G. W. Swain, M. P. Schultz. "The testing and evaluation of non-toxic antifouling coatings" *Biofouling* 10 (1996): 187.
[23] K. J. Wynne, W. Zhang, *US Patent*, 2013.
[24] E. W. Brouwers, A. A. Peterson, J. L. Palacios, L. R. Centolanza, in *American Helicopter Society 67th Annual Forum*, Virginia Beach, VA 2011.
[25] E. A. Valaker, S. Armada, S. Wilson. "Droplet Erosion Protection Coatings for Offshore Wind Turbine Blades" *Energy Procedia* 80 (2015): 263.

Diurnal evolution of wind structure and data availability measured by the DOE prototype radar system

Brian D Hirth[1], John L Schroeder[2] and Jerry G Guynes[1]

[1] National Wind Institute, Texas Tech University, Lubbock, TX USA
[2] Atmospheric Science Group, Texas Tech University, Lubbock, TX USA

E-mail: brian.hirth@ttu.edu

Abstract. A new Doppler radar prototype has been developed and deployed at Texas Tech University with a focus on enhancing the technologies' capability to contribute to wind plant relevant complex flow measurements. In particular, improvements in data availability, total data coverage, and autonomous operation were targeted to enable contributions to a wider range of wind energy applications. Doppler radar offers rapid scan speeds, extended maximum range and excellent along-beam range resolution allowing for the simultaneous measurement of various wind phenomena ranging from regional and wind plant scales to inflow and wake flow assessment for an individual turbine. Data examples and performance improvements relative to a previous edition of the technology are presented, including insights into the influence of diurnal atmospheric stability evolution of wind structure and system performance.

1. Introduction

Scanning Doppler radar has emerged as a proficient remote sensing technology for measuring wind plant complex flows. Doppler radar offers rapid scan speeds, extended maximum range and excellent along-beam range resolution allowing for the simultaneous measurement of various wind phenomena ranging from regional and wind plant scales to inflow and wake flow assessment for an individual turbine [1]. It is particularly well suited to document the spatial and temporal evolution of wind structure (e.g. gusts, lulls, wind streaks, etc.) over measurement domains comprising hundreds to thousands of square kilometers, which provides an opportunity to effectively investigate the influence of atmospheric stability on the wind flows within and surrounding wind plants [2] [3].

Historically, Texas Tech University (TTU) employed two mobile Ka-band research radars to measure wind plant complex flows on a case study basis [4] [5]. These TTUKa radar systems were originally designed for atmospheric science research applications generally focused on documenting storm scale structure. However, the radar technology and analysis techniques were successfully adapted and applied to document smaller scales of motion across wind turbine arrays.

Building on the success of the original Ka-band radar case studies, TTU designed and built a next-generation radar prototype specifically dedicated to the measurement of wind plant complex flows. In particular, this United States Department of Energy (DOE) funded radar prototype was designed to improve data availability, enhance maximum range, and provide more autonomous system operation relative to the Ka-band radars. The DOE radar prototype was successfully deployed at the Reese Technology Center (RTC) in Lubbock, TX in May 2016.

This paper will present initial analysis of wind flow measurements collected by the DOE radar prototype during the summer months of 2017. Data availability is quantitatively assessed as a function of radar range and compared to a Ka-band radar to assess data availability improvements. Documentation of the diurnal variability of the measured wind flow structure is also qualitatively assessed in the context of the potential impacts of stability on data availability and ultimately wind plant performance.

2. Technical Specifications and Data Collection Methods

The DOE radar system has been designed to run autonomously and is constructed in a modular fashion to support portability to different deployment locations (Figure 1). All core components of the radar system (transmitter, receiver chain, signal processor, operations server, data storage server, UPS, etc.) are housed within a 20-ft (6.1 m) sea container that is affixed to concrete pilings using tandem lock anchors at the four corners of the container. A secondary wall has been constructed inside the sea container which contains a door, a commercial air conditioning unit, and appropriate electrical service connections and distribution points. When the radar is deployed and operational, this wall becomes the exterior wall while the doors of the sea container are fixed in the open position. The doors are then closed for system transport. A steel exoskeleton support structure straddles the sea container and provides a platform for the radar pedestal and antenna systems to be mounted. The pedestal and antenna systems are housed within a protective radome which provides shelter to wind loading and other atmospheric hazards such as hail. Total system height from the ground to the top of the radome is approximately 8.5 m.

Figure 1. Co-located DOE-X prototype (left) and TTU Ka-band (right) radars positioned for comparative performance testing. The DOE-X radar sea container and interior wall can be seen as well as the exoskeleton support structure straddling the sea container and radome housing the pedestal and antenna systems.

The DOE radar operates in the X-band at a frequency of 9.3 GHz. The radar maintains a half-power beam width of 0.5° and an along beam resolution of 9-15 m, depending on the pulse configuration used. Variable scan speeds in excess of 30° s^{-1} can also be used. The transmitter allows a maximum pulse repetition frequency (PRF) of up to 4000 Hz for a 20 μsec compressed pulse, which yields a maximum range of 35 km. Pulse types, scan speeds and PRFs can all be configured to accommodate the data collection objective of interest.

Data were collected between 30 June and 28 September 2017 at the RTC to assess system data availability spanning several months. The radar system was configured to run a single 360° horizontal

Plan Position Indicator (PPI) scan at the 1.0° elevation tilt every five minutes. Two scans speeds were used, 2.5° s^{-1} and 30° s^{-1} to assess the impact of varying scan speed (thus varying the number of available samples to construct the Doppler velocity spectra) on data availability. Scan durations were 144 s and 12 s, respectively. A 20 μs frequency modulated compressed pulse was used, resulting in an along-beam range resolution of 15 m between the first available range bin at 3 km and the maximum range bin of 32 km. Specifications for the DOE radar operation during this data availability study are provided in Table 1. In total 23,955 individual PPI scans for each scan speed were acquired during the approximately three month period. Minimal "forced" radar downtime occurred during several short time periods to accommodate other nearby experiments at RTC requiring radar silence.

Table 1. Technical specifications of the DOE prototype radar used for the summer 2017 RTC data availability study.

Parameter	Specification
Peak Transmit Power	11 kW
Transmit Frequency	9.3 GHz
Wavelength	3 cm
Antenna Diameter	4.6 m
Half-Power Beam Width	0.5°
Pulse Length	20 μs
Range Gate Spacing	15 m
PRF	4000 Hz
Range	3-32 km
Azimuthal (PPI) Resolution	0.352°
Pointing Accuracy	0.05°
Velocity Accuracy	0.03 m s^{-1}
Horizontal Scan Speed	2.5, 30° s^{-1}

Radar data were collected through a wide range of atmospheric conditions including both precipitating and non-precipitating (e.g. clear-air) environments. A simple quality control procedure was implented on the raw radial velocity data to remove the influence of ground clutter and point targets, second-trip echos (range aliasing) and noise. Signficant diurnally driven variability in the structure of the radial velocity field was common, particularly in clear-air enviroments where minimal sensible weather was occuring.

3. Diurnal Variability

Diurnal trends through the day of 06 September are presented in Figure 2. PPI scans using the 30° s^{-1} scan speed are shown at 0301 UTC, 0901 UTC, 1801 UTC and 2101 UTC (Figure 2a-d). A 5-min average potential temperature (Figure 2e) and turbulence intensity (Figure 2f) profile is constructed using measurments from nine instrumentation levels (between 1-158 m) of the TTU meteorological tower located roughly 345 m southwest of the DOE radar (Figure 2e). The tower samples continuously at 50 Hz. Significant diurnal evolution of the boundary layer wind structure is evident on the presented PPI scans which can be related to background stability changes. During the evening and overnight hours (e.g. 0301 UTC and 0901 UTC), the boundary layer wind structure is qualitiatively smooth coincident with a temperature inversion noted on the meterological tower, indicative of stable atmsopheric conditions. Turbulence intensity values constructed from the meteorological tower sonic anemometer wind measurements are quite low during this same time period, generally less than 10% above 10 m. As atmospheric mixing begins to increase with surface heating, a more streaky structure is seen in the

IOP Conf. Series: Journal of Physics: Conf. Series **926** (2017) 012003 doi:10.1088/1742-6596/926/1/012003

boundary layer flow (e.g. 1801 UTC) as the temperature inversion has mixed out and the stability regime has started to transition back towards unstable. Further surface heating and mixing into the afternoon hours (e.g. 2101 UTC) results in an unstable boundary layer with much larger turbulent structures evident in the boundary layer flow. During the afternoon hours, turbulence intensity values increase to 25-40% above 10 m. This diurnal evolution is prevalent most days during the test period lending insights into the evolution of stability and turbulence in the background boundary layer flow on a day to day basis. Obtaining a better understaning of these dirunal variations is of great importance for the wind energy industry to improve turbine wake and wind plant layout modelling, as well as develop the next generation of (proactive) turbine control strategies to further optimize collective wind plant performance.

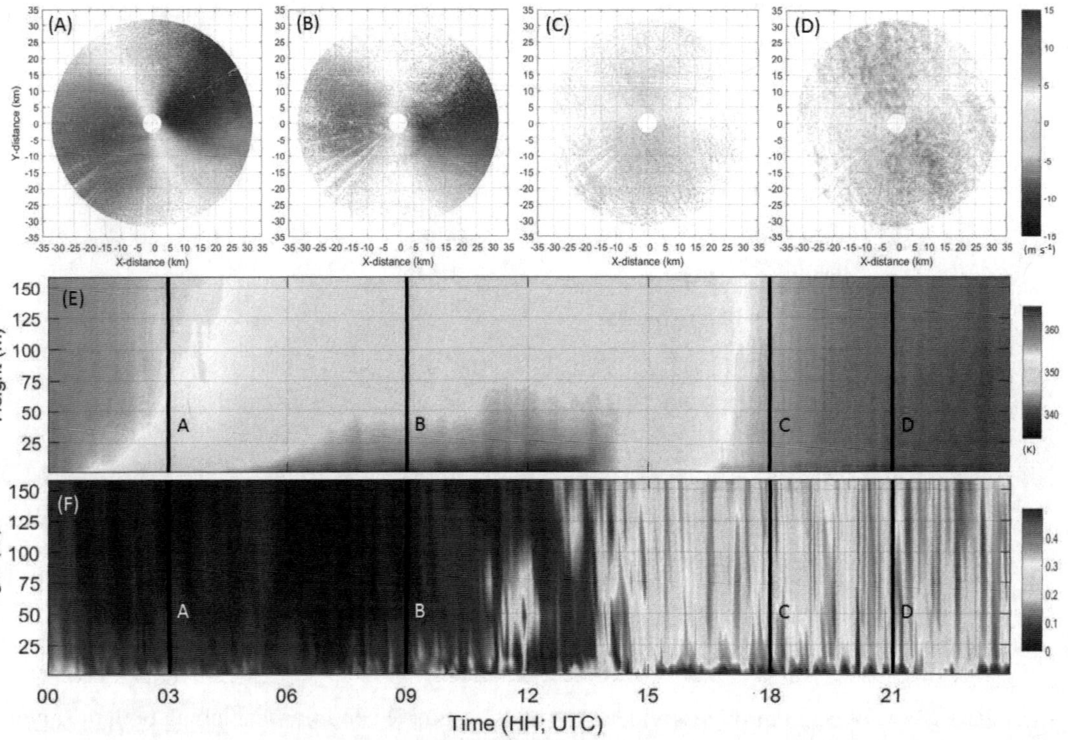

Figure 2. DOE radar measured radial velocity (m s^{-1}) PPI sweeps at the 1.0° elevation tilt using a scan speed of 30° s^{-1} from 06 September at (A) 0301 UTC, (B) 0901 UTC, (C) 1800 UTC and (D) 2100 UTC. TTU instrumented tower 5-minute (E) averaged potential temperature (K) profile and (F) turbulence intensity for the day of 06 September.

4. Data Availability Assessment

Data availability was assessed for each PPI scan as a function of radar range. Since a 360° PPI scan was used with an azimuthal sampling resolution of 0.352°, approximately 1022 data bins exist at a given range ring from the radar. Data availability for each range ring was constructed as the percentage of azimuthal bins containing valid data compared to the total available bins (Figure 3). Nominally, data availability decreased with range, though the variety of atmospheric conditions measured contributed to the construction of diverse data availability curves. For example, during a period of widespread precipitation, data availability never dropped below 94% (Figure 3a) while during a period of marginal clear-air return, data availability decreased from an average of 87% between 3-5 km to 48% between 20-22 km. Figure 4 presents the composite mean data availability analysis with range using the full set of 23,995 PPI scans. Using the 30° s^{-1} scan speed, availability at 3, 5, 10, 20 and 30 km is 72%, 70%, 58%, 35% and 18%, respectively. When the scan speed is slowed to 2.5° s^{-1}, availability increases to 84%, 81%, 70%, 47% and 27%, respectively for the same ranges. Therefore, slowing the scan speed yielded an availability increase of 12% at near ranges and 9% at far ranges for the period.

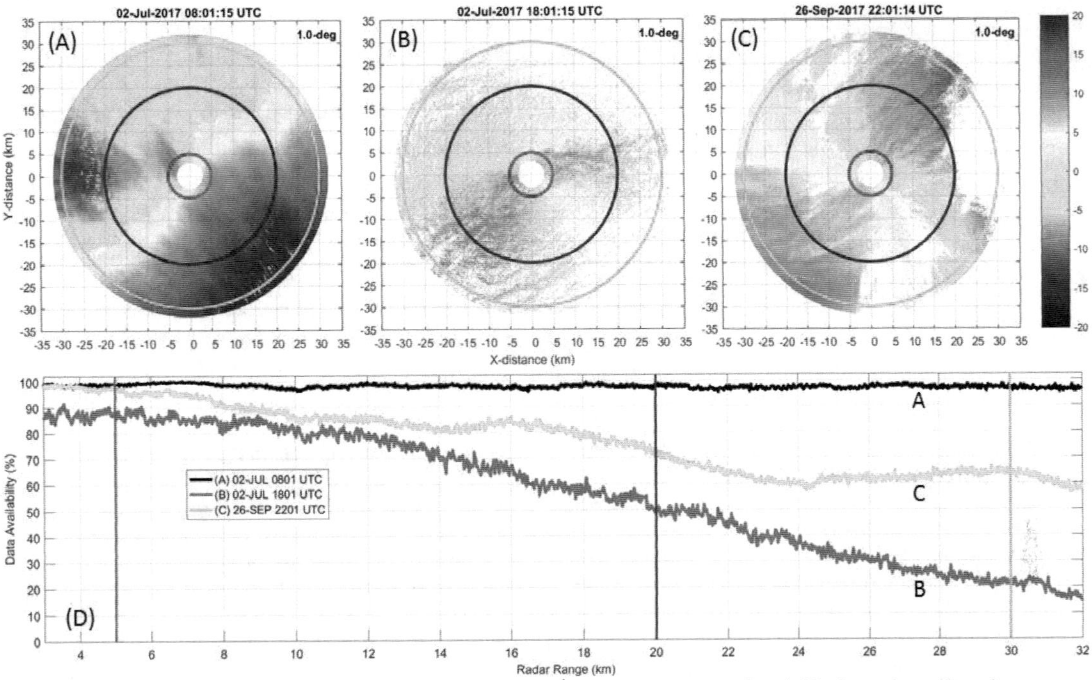

Figure 3. DOE radar measured radial velocity (m s^{-1}) PPI sweeps at the 1.0° elevation tilt using a scan speed of 30° s^{-1} from (A) 02 July at 0801 UTC, (B) 02 July at 1801 UTC and (C) 26 September at 2201 UTC. The red, blue and green circles represent the 5, 20 and 30 km range rings, respectively. (D) Data availability (%) assessment with range (km) for the individual PPI scans A-C. Red, blue and green vertical bars represent the 5, 20 and 30 km range distances, respectively.

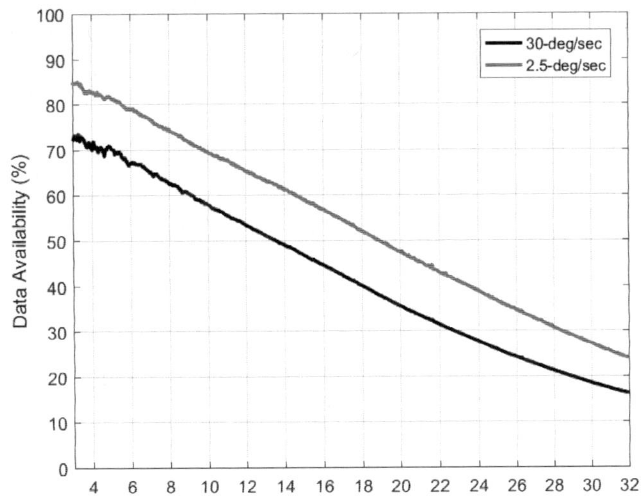

Figure 4. Composite DOE prototype radar data availability (%) assessment with radar range (km) using the 30° s^{-1} (black curve) and 2.5° s^{-1} (red curve) scan speeds.

Data availability was also assessed by time of day to note diurnal trends (Figure 5). During the data collection period, daytime hours ranged from roughly 11 UTC – 01 UTC while night time hours ranged from roughly 01 UTC – 11 UTC. When segregating and compositing the full collection of PPI scans

IOP Conf. Series: Journal of Physics: Conf. Series **926** (2017) 012003 doi:10.1088/1742-6596/926/1/012003

into hourly bins, two distinct availability maximums are evident, the first between 02-04 UTC and the second between 16-19 UTC. As discussed in Section 2, this bimodal distribution of availability appears directly linked to diurnal stability evolution and the associated scattering mechanisms associated with each. For example, during the daytime hours, it is believed that convective boundary layer mixing generated by surface heating aided in lofting and distributing effective radar scatterers (pollen, dust, etc.) supportive of higher availability. As the afternoon transitioned to evening, this mixing process weakened allowing the distributed scatterers to settle yielding a relative decrease in availability. During the evening and overnight hours, insect activity increased and surface cooling aided in the development of a boundary layer temperature inversion contributing to a pronounced vertical density gradient. It is believed the resulting vertical density gradient yields favorable conditions for the DOE prototype radar to benefit of Bragg scattering. Near and just after the time of sun rise, a general weakening in the overnight density gradient created a decrease in availability. As the sun angle continued to increase during the morning hours, surface heating once again contributed to the start of surface mixing, and the cycle repeated. Based on measurements using the 30° s^{-1} scan speed, the net result of this diurnal cycle was a peak in composite availability at 3 km range of 72% during the day time hours and 81% during the night time hours with an associated minimum in availability just after sunrise of 58%.

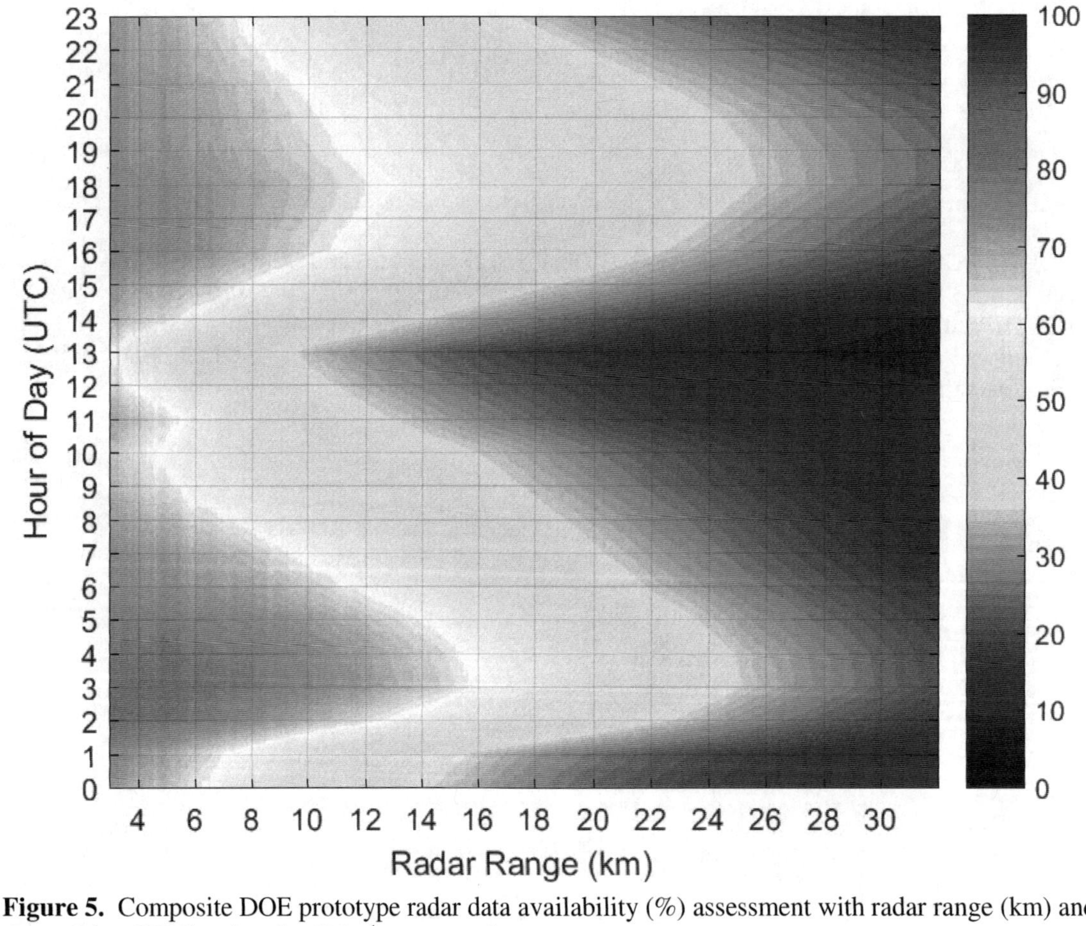

Figure 5. Composite DOE prototype radar data availability (%) assessment with radar range (km) and time of day (UTC) using the 30° s^{-1} scan speed.

An important distinction between the DOE prototype radar and other currently available scanning remote sensing technologies is the significant maximum range available, and therefore extensive domain of possible data coverage. If data were 100% available between 3-32 km, the total spatial coverage of radar measurements would be nearly 3200 km^2. To assess the impact of data availability on total data coverage, the summed area of all valid data on each individual PPI sweep using the 30° s^{-1} scan speed (e.g. scan duration of 12 s) is calculated and compared to the data availability assessed at the 3 km range (Figure 6). When data availability at 3 km is in excess of 90%, mean total data coverage across the entire PPI sweep is 1993 km^2. Availability at 3 km of 70-80% yields a mean areal coverage of 1037 km^2 across the entire PPI sweep and between 50-60% yielded a mean areal coverage of 592 km^2 across the entire PPI sweep.

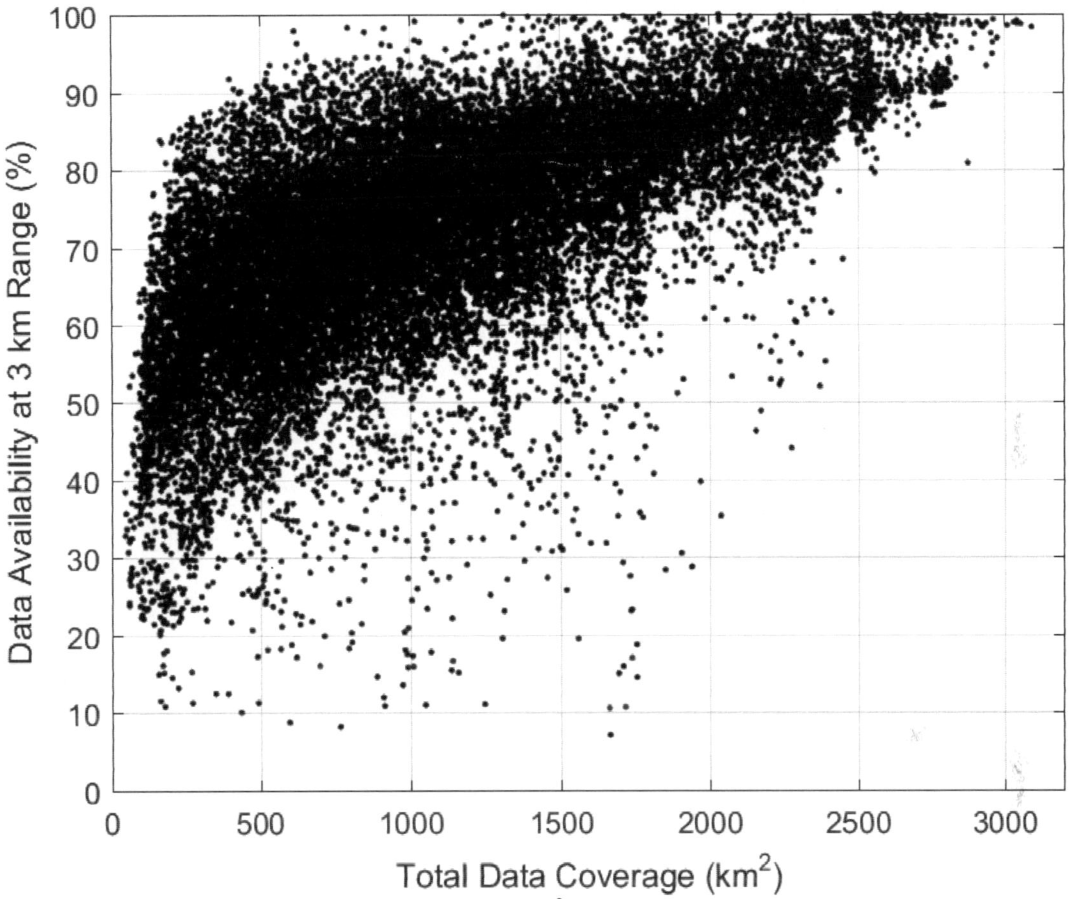

Figure 6. DOE prototype radar data coverage (km^2) versus availability (%) at 3 km range for each individual PPI sweep using the 30° s^{-1} scan speed.

5. Performance Comparison to TTUKa

To better understand the magnitude of data availability enhancement achieved between the existing TTUKa radars and the DOE prototype radar, a single TTUKa radar was deployed beside the DOE system during a period of relatively low data availability on 3-4 August 2016 (Figure 1). Each radar scanned an identical 90° sector at the 0.5° elevation tilt once every 15 minutes for a total of 97 scans. Data availability was again assessed as a function of range using the methods described above, where the maximum range of the TTUKa radar was 10 km (using the configuration settings for a nominal wind plant data collection campaign). In general, the DOE prototype radar demonstrated a significant increase in data availability at all times and at all ranges, with mean increases in availability of 38%, 38%, 35% and 32% at 4, 6, 8 and 10 km range, respectively (Figure 7), and a peak availability increase of 75%. A significant finding from this comparison was the substantial increase in availability found during the night time hours, with increases between 60-70% common at all four ranges assessed. In fact, beyond 4-6 km, little to no data availability can even be found in the TTUKa radar measurements, which may be due to the smaller wavelength of the TTUKa radar being unable to benefit from Bragg scattering.

IOP Conf. Series: Journal of Physics: Conf. Series **926** (2017) 012003 doi:10.1088/1742-6596/926/1/012003

These initial comparisons show that the DOE prototype system represents a monumental advancement in data availability capability from the existing TTUKa radar technology. The area of potential data coverage is also substantially increased in the DOE prototype system compared to the existing TTUKa radar when both system are configured to run for a nominal wind plant measurement campaign. As previously discussed, the available data range for the DOE system is 3-32 km while for the TTUKa system is 3-10 km, both with an along beam resolution of 15 m. The resulting total areal coverage of the DOE radar is 3200 km^2 as compared to only 285 km^2 for the TTUKa system, a factor increase of greater than 11.

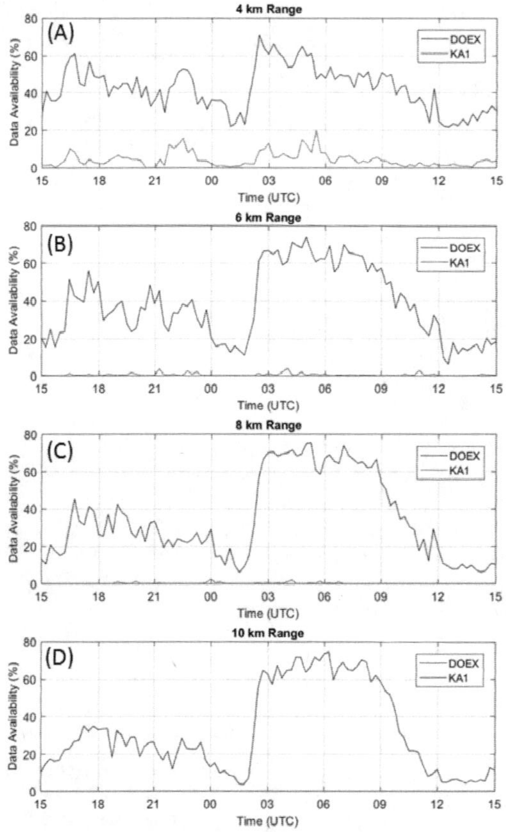

Figure 7. Data availability (%) comparison between the DOE prototype radar and a TTUKa radar on 3-4 August 2016 for a 24 hour period at the ranges of (A) 4 km, (B) 6 km, (C) 8 km, and (D) 10 km. The DOE radar availability is denoted by the black line and TTUKa availability by the red line.

6. Concluding Thoughts

The DOE prototype radar was constructed to provide increased data availability, enhanced maximum range and autonomous operation when compared to the existing TTUKa radars. Deployed at the RTC in Lubbock, TX a data availability study was conducted during the summer months (30 June – 29 September) of 2017. Data were collected using a 360° PPI scan on a 1.0° elevation tilts using the scan speeds of 2.5° s^{-1} and 30° s^{-1}. Statistics generated from measurements obtained using the faster scan speed and collected in a variety of atmospheric conditions reveal a composite data availability of 72% at 3 km range, decreasing to 58% and 35% at 10 and 20 km range, respectively. Using the slower scan speed, data availability at 3 km range is 84%, an increase of 12%. When compared to the existing TTUKa radar, data availability is markedly increased on the order of 40-70%, especially during the nocturnal hours where the DOE prototype radar produced considerably higher quality data when compared to the TTUKa availability. The total area of potential data coverage provided by the DOE prototype radar is also greater than 11 times that which was historically provided by the TTUKa systems.

Hourly segregated and composited data availability highlighted a clear diurnal trend, with relative maximums in the afternoon and nocturnal hours, and relative minimums in the morning and early evening hours. This evolution of availability appears tied to diurnal cycles of boundary layer stability and the subsequent influence of stability on the existence of relevant scatterers supportive of quality measurements by the DOE prototype radar. Background turbulence was qualitatively assessed by viewing individual PPI scans and qualitatively supported by data collected from a nearby meteorological tower instrumented at multiple levels during a single diurnal cycle. Smooth boundary layer flow was observed during periods characterized by a temperature inversion and low turbulence intensity (less than 10% above 10 m) while larger turbulent structures were observed during periods where the temperature profile was more uniform as a result of boundary layer mixing and turbulence intensity was generally much higher (25-40% above 10 m).

The DOE prototype radar developed at TTU represents a substantial advancement in the effort to employ Doppler radar to measure wind plant complex flows. Data availability and maximum range enhancements allow for wind measurements to reliably be made in a much wider spectrum of atmospheric conditions and over larger spatial domains compared to the original TTUKa radars. In particular, the DOE prototype allows for the investigation of wind plant complex flows throughout the diurnal cycle. These results show that the DOE prototype radar is well positioned to assist the wind energy community in enhancing its understanding of wind plant complex flows. The first commercial application [1] [6] of this technology is currently being evaluated for offshore wind plant monitoring at the Westermost Rough wind plant in the UK.

7. References

[1] Hirth B.D., Brink F., Schroeder J.L., Guynes J.G., Nygaard N., Gretlund J. (2017) *A dual-Doppler radar system to document wind plant complex flows.* Offshore Wind Energy 2017 Conf. Proc., London, England.

[2] Wharton S., Lundquist J.K. (2012) *Assessing atmospheric stability and its impacts on rotor-disk wind characteristics at an onshore wind farm.* Wind Energy, 15, pp. 525-546.

[3] Machefaux, E., Larsen G.C., Koblitz T., Troldborg N., Kelly M.C., Chougule A., Hansen K.S., Rodrigo J.S. (2016) *An experimental and numerical study of the atmospheric stability impact on wind turbine wakes.* Wind Energy, 19, pp. 1785-1805.

[4] Hirth B.D., Schroeder J.L., Gunter W.S., Guynes J.G. (2015) *Coupling Doppler radar-derived wind maps with operational turbine data to document wind farm complex flows.* Wind Energy, 18, pp. 529-540.

[5] Hirth B.D., Schroeder J.L., Irons Z., Walter K. (2015) *Dual-Doppler measurements of a wind ramp event at an Oklahoma wind plant.* Wind Energy, 19, pp. 953-962.

[6] Nygaard N., Brink F., Gretlund J., Hirth B.D., Schroeder J.L., Guynes J.G. *Wake measurements in an offshore wind farm using dual-Doppler radars.* Wind Energy Science Conference 2017 Conf. Proc., Lyngby, Denmark.

Acknowledgments
The DOE prototype radar was developed with funding provided by the United States Department of Energy under Award No. DE-EE0006804.

IOP Conf. Series: Journal of Physics: Conf. Series **926** (2017) 012004 doi:10.1088/1742-6596/926/1/012004

Using Sentinel-1 SAR satellites to map wind speed variation across offshore wind farm clusters

S F James

STFC Rutherford Appleton Laboratory, Harwell Campus, Didcot, OX11 0QX, UK.

sarah.james@stfc.ac.uk

Abstract. Offshore wind speed maps at 500m resolution are derived from freely available satellite Synthetic Aperture Radar (SAR) data. The method for processing many SAR images to derive wind speed maps is described in full. The results are tested against coincident offshore mast data. Example wind speed maps for the UK Thames Estuary offshore wind farm cluster are presented.

1. Introduction

This paper demonstrates how high resolution maps suitable for studying wind flow and wakes around offshore wind farms can be derived from open access satellite data. It presents an automatic method for processing Sentinel-1 Synthetic Aperture Radar (SAR) data [1] over many images that reduces the impact of the turbines' own radar reflections and calculates wind speed values every five hundred metres across scenes of tens to hundreds of kilometres. The results are tested against coincident offshore mast data for 25 SAR scenes. The purpose is to develop and share a technique to make the large amount of routine, free and open access Sentinel-1 SAR data available to advance understanding of both wind flow around offshore wind farms and the impact of wind farms on the flow.

Synthetic Aperture Radar (SAR) is an active microwave radar instrument. That is it transmits and receives. It does not rely on illumination by daylight, and at the wavelengths used (of the order of centimetres) it isn't impacted by cloud cover so it can operate at any time of day. The SAR beam interacts with the fine, centimetre scale surface roughness of water to affect the scattered return radar signal received by the SAR instrument. Wind speed across a water surface affects the degree of roughness and hence the received radar signal. This is the basis for measuring offshore wind speeds with SAR. Empirically derived Geophysical Model Functions (GMF) are used to relate the calibrated radar return to the wind speed at a height of 10m. GMFs usually relate the radar backscatter to wind speed and wind direction, so to find wind speed, wind direction data is a necessary input and this can be derived from image features or taken from other data.

A number of other features can also affect the radar return from the sea surface including surface currents, rain cells, oil spills and sometimes bathymetry in shallow water: SAR cannot make a direct measurement of wind speed at the sea surface and uncertainty arises in deriving the wind speed from the radar return. SAR, in common with other coherent radar instruments, features speckle noise which gives SAR images a grainy, salt and pepper look. This is because the signal returned from any smallest resolution cell will be a vector sum of the return from the scatterers in that cell and will thus vary from cell to cell randomly. In practice, this limits the resolution possible for wind speed results from a SAR image, as some averaging across pixels must take place to reduce the random noise effect of speckle, leading to a trade-off between noise reduction and resolution.

Offshore wind speed retrieval from SAR is used in a wide range of fields from the study of mesoscale wind phenomena such as cyclones and atmospheric gravity waves, to wind energy [2-7]. A number of studies have specifically applied it to looking at external wakes from offshore wind farms [8-11]. These use a number of methods to assess the velocity deficit seen behind the wind farm: it is not straight forward to decide where to measure the reference, undisturbed wind speed as there are often underlying wind speed changes across SAR scenes, such as offshore and onshore wind speed gradients. Hasager et al [10] have developed a method to combine results from many SAR scenes to overcome this issue and this type of approach lends itself to the increased availability of SAR data arising from the Sentinel-1 mission.

The Sentinel-1 mission of two satellites, A and B, each carrying a SAR instrument is potentially significant for offshore wind studies because it provides free and open access SAR data. There is routine coverage especially over Europe. The data provided are recent and ongoing giving a good time overlap with many operating offshore wind farms. Previously SAR data have typically been expensive, or available in small amounts to academics, or available from older missions when far fewer offshore wind farms were operating. Sentinel-1A provides regularly available data following a calibration enhancement (https://sentinel.esa.int/documents/247904/2142675/Sentinel-1A_TOPS_Radiometric_Calibration_Refinement), from 25th November 2015. Sentinel-1B was launched later and has been providing regular data since late September 2016. The times at which data are available are limited. Across Europe the two satellites provide total coverage every 6 days. At any one location, images from both 1A and 1B will usually be taken at around the same time of day on the ascending satellite passes and at around the same time of day on the descending passes. (During an ascending pass a satellite passes over Europe approximately south to north, and on a descending pass approximately north to south.) So for example, images covering the Thames Estuary, UK are generally all at around 06:00 UT on descending passes or 17:40 UT on ascending passes.

There is also free software available to handle Sentinel data: the Sentinel Application Platform (SNAP). SNAP has a graphical user interface (GUI) and a command line interface. The command line interface in particular allows the automation of many tasks. The ability to derive wind fields from SAR data with a largely automated process allows the use of multiple scenes to provide climatological results about impacts of offshore wind farms on the wind flow under different conditions.

2. Methodology

2.1. Overview

The approach taken here has been to make as much use as possible of the tools made available with the Sentinel data and to extend these where necessary to produce results suitable for understanding wind fields around offshore wind farms at useful resolutions. In addition, the derivation of wind speeds should be as automated as possible, so that numbers of SAR scenes can be compared. In order to verify that meaningful wind speeds can result from this method, results are compared to coincident high quality, high time resolution mast data from the Met Mast Ijmuiden (MMIJ). The method for calculating wind speeds from SAR data is summarised in the flow chart in Figure 1.

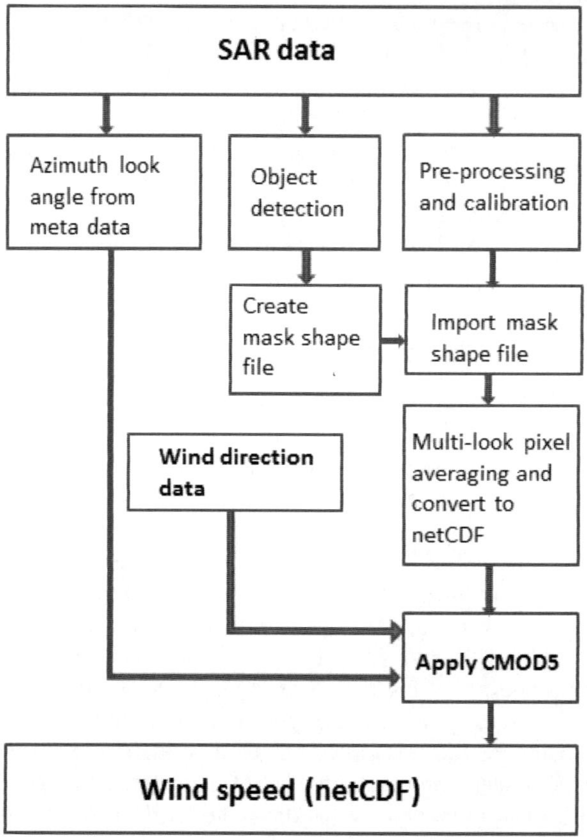

Figure 1. Flow chart for calculating wind speeds from SAR

2.2. Pre-processing and calibration

The Sentinel Application Platform (SNAP) provides the tools needed to calibrate Level 1 (L1) ground range detected (GRD) Sentinel-1 data products to get the Normalised Radar Cross Section (NRCS) values, known as Sigma0, which are then used to calculate the wind speed. The SNAP pre-processing steps used on the L1-GRD data are orbit correction, thermal noise removal, radiometric calibration (to calculate Sigma0), application of a land mask and multi-looking to smooth out speckle noise inherent in the radar data.

2.3. Masking bright objects

Wind turbines and ships are strong radar reflectors and so present as very bright objects in a SAR offshore scene. If they are not masked, they will distort the derived wind speed. The SNAP Ocean Object Detection tool is used to identify bright objects automatically. A minimum target size of 10m and a maximum target size of 470m were used. This is a tool primarily aimed at identifying shipping [12]. It produces an xml report which gives the location and estimated size of detected objects. SNAP doesn't currently provide a method to create a mask from the Object Detection Tool results, except manually in the GUI. In order to be able to automate the data processing, Python code was written to parse the xml report; to create buffered point shapes at each object, and to use those to create a single, union shape file which could then be imported as a mask and used in subsequent processing. The Python modules Shapely and Fiona made this quite straight forward. It was decided to make the size of each buffered point the larger of 100m or the detected object size to be more certain of fully masking turbines.

2.4. Wind field estimation

SNAP also provides a wind field estimation tool. This uses a Fast Fourier Transform (FFT) method to determine wind direction from image features and then implements the geophysical model function CMOD5 [13] using these direction data. The FFT method was observed to fail at resolutions useful to this study, a known limitation of the method [14]. The default grid size in the SNAP wind field estimation tool is 20km by 20km, but the aim of this work was to be able to create wind speed maps at resolutions of 1km or less. As a result, CMOD5 was implemented in Python to allow the introduction of alternative wind direction information. The CMOD5 function relates Sigma0 from co-polarised (vertically polarised transmit and receive known as VV) C-band SAR data to wind speed, wind direction and SAR incidence angle. A look up table is created by using CMOD5 to calculate Sigma0 for a range of wind speeds, relative wind directions and incidence angles. This look up table can then be used to find the corresponding wind speed for a given Sigma0 measurement, relative wind direction and incidence angle. The relative wind direction is the wind direction relative to the satellite azimuth look angle.

A look up table of Sigma0 values derived from the CMOD5 geophysical model function (GMF) was created with a value for Sigma0 for all combinations of wind speeds from 1 to 60 ms^{-1} in 0.2 ms^{-1} bins, incidence angles from 20° to 46° in 0.2° bins, and relative wind directions from -180.0° to 180.0° in 2.5° bins.

2.5. Other inputs to CMOD5

The azimuth look angle of the SAR image is needed to find the relative wind direction. This has been calculated from the corner coordinates of each image. These are found in the .manifest meta data file that comes with each Sentinel-1 data file. The look direction is given by the bearing from the 1st to the 2nd or from the 4th to the 3rd corner coordinates. Python code made available by Jérôme Renard at https://gist.github.com/jeromer/2005586 was used to find a bearing between two points.

The remaining input data necessary for calculating wind speed is the wind direction. This study uses global reanalysis data MERRA2 [15] to provide wind direction. Global reanalysis data provide standardised meteorological and atmospheric data from forecast models run after the event. MERRA2 is available at time resolutions of up to one hour and spatial resolution of 0.5° latitude by 0.625° longitude. MERRA2 includes 1-hourly eastward and northward wind components at 2m, 10m and 50m in the 2d, 1-Hourly, Time-Averaged, Single-Level, Assimilation, Single-Level Diagnostics product (tavg1_2d_slv_Nx). As the CMOD5 GMF relates Sigma0 to winds at 10m, hourly MERRA2 10m wind components were used to give an input wind direction.

2.6. Applying CMOD5

The command line version of SNAP was used to carry out the following steps on level 1, ground-range-detected (GRD) Sentinel-1 data VV band:

1. Apply orbit corrections.
2. Remove thermal noise.
3. Radiometric calibration to give Sigma0.
4. Apply a land mask.
5. Import as a vector the shape file created as a result of object detection.
6. Apply this vector as a mask.
7. Reduce speckle noise by averaging Sigma0 over 500m squares by applying 50 by 50 multi-looking.
8. Export the results in netCDF4 format, selecting to save parameters Sigma0 and incidence angle.

The netCDF4 format is designed to support array-oriented scientific data. The netCDF4 results contain a grid of data with x, y coordinates and for each x, y point a value for the latitude, the longitude, the incidence angle and Sigma0. The relative wind direction is given by the local wind direction relative to north minus the satellite azimuth look angle relative to north. For each grid point in the input data, the CMOD5 lookup table created earlier is used to look up the wind speed corresponding to Sigma0, the incidence angle and the relative wind direction.

2.7. Verifying the wind speed results

Although the approach of this work is to seek to understand the spatial variation, rather than to find absolute wind speeds, it is clearly necessary to verify that the wind speeds calculated are related to reality, or the variability may be spurious too.

High quality, very high time resolution mast data have been obtained from Met Mast Ijmuiden (MMIJ) for this purpose. MMIJ is a Dutch offshore mast located well offshore, between the Netherlands and the UK. The location is 52.84817° latitude, 3.43567° longitude. Wind speed and direction data are measured at three locations equally spaced around the mast at heights of 27m and 58m. Wind speed is measured on two sides of the mast and wind direction is measured at three equally spaced locations at 90m, which is the top of the mast. Air temperature data are available at heights 21m and 90m. Wind speed, wind direction and air temperature are all available at 4Hz time resolution. Sentinel-1 data are available from 25th November 2015 and MMIJ mast data is available until 11th March 2016. Coincident data from the mast and Sentinel-1A are available between November 2015 and March 2016 for 25 images.

The CMOD5 GMF relates the Sigma0 measured by the radar to the wind speed at a height of 10m. The mast wind speed data are processed to provide wind speeds at 10m for comparison. At 27m a 'true' wind direction is calculated as follows, in order to minimize mast shading effects. The directions around the mast are considered as six equal sectors, with a wind vane in the centre of every other sector. There are three measured wind directions: one will be the lowest value, one the highest and one in the middle. The middle value of the three measured wind directions is identified and the sector containing this wind direction is noted. One wind vane will be located in this sector, or in the sector directly opposite. This wind vane is disregarded as likely to be affected by mast shading, and the true wind direction measurement is taken to be the average of the values measured at the other two wind vanes. At 27m a 'true' wind speed is calculated by averaging the wind speed measurement from the two anemometers closest to the true wind speed direction at 27m. At 90m there are only two cup anemometers and this is at the top of the mast, so the wind speed is taken to be the average of the two measurements. The wind speed at 27m is adjusted to 10m using a Bulk Richardson method to determine stability conditions, which makes use of the mast air temperature data at 21m and 90m and the wind speed at 27m and 90m.

It is necessary to compare time series mast data at one location with SAR wind speeds located across a scene measured in a very short time period (less than half a minute). The 10m mast wind speeds are averaged over 10 minutes, for a time period centred on the mid-point of the SAR image time. The 10 minute average wind speed is chosen as it is a widely available standard in wind condition measurements. Assuming that the wind flow holds to Taylor's frozen turbulence hypothesis, the spatial distance which should be compared to this time period is the distance crossed at the 10 minute average wind speed in 10 minutes. So for each SAR image, the SAR wind speeds are averaged across a footprint centred on the mast and with a diameter equal to the distance covered by the average mast wind speed in 10 minutes [11]. The SAR footprint average wind speed is compared to the mast 10 minute average wind speed.

3. Results and Discussion

3.1. Verifying the wind speed results at Met Mast IJmuiden

The wind speed was derived from 25 SAR scenes coincident with data available from the MMIJ offshore mast. The wind speed at the mast was adjusted to a height of 10m, using air temperature data to account for atmospheric stability conditions. The mast 10m wind speeds were averaged over 10 minutes with the 10 minutes centred on the time of the SAR image. The distance covered by the 10 minute average wind speed measured by the mast in those 10 minutes was used as the diameter of a footprint centred on the mast. The SAR derived wind speed was averaged for all points falling in this footprint. The results are shown in Table 1.

Table 1. Summary results of wind speeds at MMIJ

	SAR image start (YYYY-MM-DD HH:MM:SS)	SAR image end (HH:MM:SS)	MMIJ 10m wind speed, 10 minute average (ms^{-1})	Footprint radius (m)	SAR 10m wind speed, footprint average (ms^{-1})
1	2015-11-28 17:33:03	17:33:25	17.61	5283	19.17
2	2015-12-07 05:57:32	05:58:01	5.11	1532	4.74
3	2015-12-10 17:33:03	17:33:24	16.00	4800	16.20
4	2015-12-12 06:05:58	06:06:23	9.51	2854	10.88
5	2015-12-19 05:57:43	05:58:12	11.77	3530	8.81
6	2015-12-22 17:33:02	17:33:24	14.79	4437	15.50
7	2015-12-24 06:05:57	06:06:22	14.55	4364	12.89
9	2016-12-31 05:57:31	05:58:00	14.87	4460	15.42
10	2016-01-03 17:33:02	17:33:23	15.53	4658	17.17
11	2016-01-05 06:05:57	06:06:22	10.10	3029	9.32
12	2016-01-12 05:57:42	05:58:11	10.43	3130	12.19
13	2016-01-15 17:33:01	17:33:23	11.10	3330	11.24
14	2016-01-17 06:05:56	06:06:21	7.07	2121	7.82
15	2016-01-24 05:57:31	05:58:00	11.41	3422	9.37
16	2016-01-27 17:33:01	17:33:22	17.11	5134	16.19
17	2016-01-29 06:05:56	06:06:21	17.24	5173	14.73
18	2016-02-05 05:57:42	05:58:11	10.37	3112	11.10
19	2016-02-08 17:33:01	17:33:22	19.14	5743	21.27
20	2016-02-10 06:05:56	06:06:21	10.74	3223	12.28
21	2016-02-17 05:57:30	05:57:59	9.00	2701	8.77
22	2016-02-20 17:33:01	17:33:22	12.99	3896	13.78
23	2016-02-29 05:57:41	05:58:10	7.73	2319	7.66
24	2016-03-03 17:33:01	17:33:22	4.54	1361	6.70
25	2016-03-05 06:05:56	06:06:21	6.88	2064	7.93

The results of plotting the 10 minute average mast wind speed against the equivalent footprint averaged SAR wind speed for all available scenes is shown in Figure 2.

The plot shows a linear correlation, with a Pearson's r value of 0.93 indicating a strong, linear correlation between the wind speed derived from the SAR images and the wind speed measured at the mast. The r^2 measure of 0.87 indicates that the amount of variance in wind speed derived from the SAR images that is explained by the wind speed measured at the mast is 87%.

As CMOD5 relates the calibrated radar signal of Sigma0 to wind speed and wind direction, it is reasonable to consider whether the use of spatially and temporally imprecise MERRA2 wind direction data is having a significant impact on the quality of results. In this case we have on site wind direction measurements available from the mast. The results of using the mast measured wind direction as an input in deriving wind speed from SAR are shown in Figure 3. Again the 10 minute average wind speed at the mast is used and the SAR wind speed over an equivalent spatial footprint.

There is an improvement in the linear correlation, with Pearson's r now equal to 0.96 and r^2 equal to 0.92. It is clear though that the difference between the SAR derived wind speed and the wind speed measured by the mast is not solely or even largely due to poor wind direction input data. This is a useful result as generally in applying this technique to studying wind fields around wind farms coincident mast measurements will not be available, but MERRA2 will be.

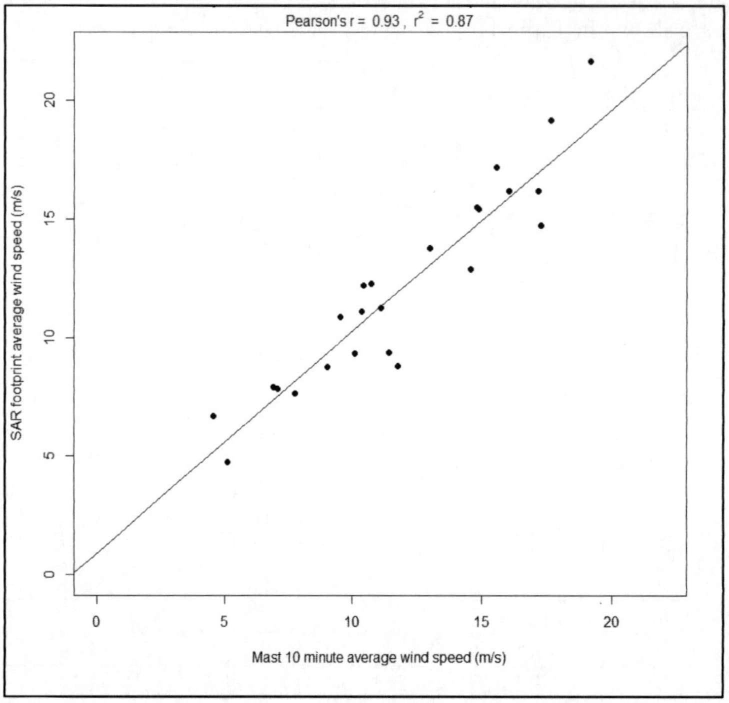

Figure 2. Comparison of mast and SAR wind speeds at MMIJ

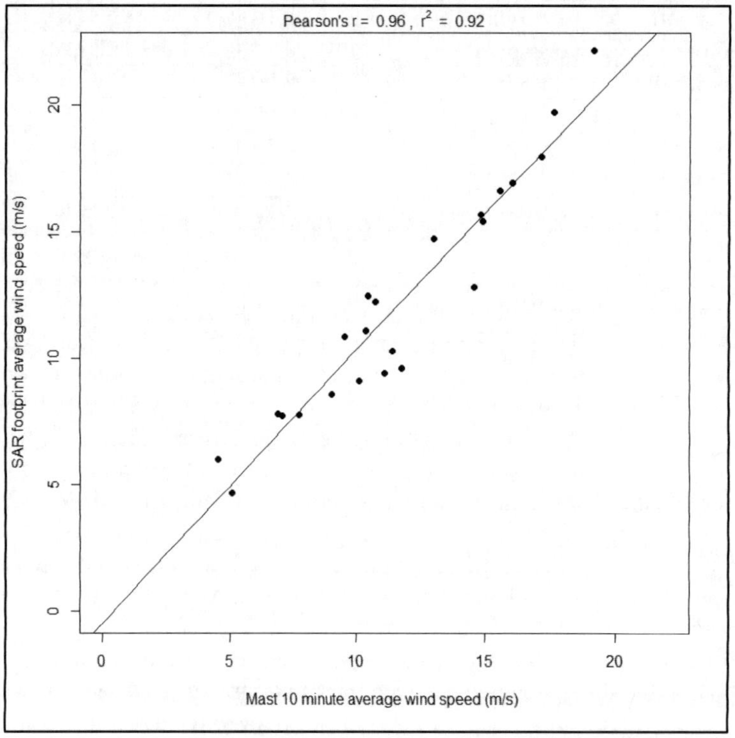

Figure 3. Comparison of mast and SAR wind speeds at MMIJ with SAR input wind direction from the mast

3.2. Further work to test and improve the method

There is scope to try different wind direction data inputs with this method, for example by using the SNAP wind field estimation tool to calculate a wind direction on a larger grid. There is also scope to test the quality of the wind speed results for coarser and finer resolutions of calibrated Sigma0 input. The resolution of the calibrated radar signal depends on the level of multi-looking to reduce speckle noise, where increased multi-looking improves speckle noise but reduces spatial resolution. Further work to test this method against different wind direction inputs and for different levels of multi-looking would be useful to extend understanding of the limitations and accuracy available.

It would also be very desirable to test the method against longer time periods of offshore mast data and in different locations. The constraint here is the availability of such data in time periods that overlap Sentinel-1 operations.

3.3. Application to offshore wind farm clusters

The purpose of this work has been to provide wind field data for further investigation of the external wakes from offshore wind farms in clusters. We are interested in understanding the scale and likelihood of external wakes impacting on neighbours in clusters. As the industry expands, offshore wind farms are often located in clusters of several wind farms together in the same sea area. External wakes, that is wakes due to whole wind farms, may impact on other, nearby wind farms within a cluster. Better understanding of how often external wakes occur and with what extent and velocity deficit can inform decisions on the siting wind farms in relation to one another and may be expected to improve the accuracy of resource prediction.

Using the method described here the wind speed maps in figures 4 to 9 have been produced. They all show an area of the Thames Estuary with a number of wind farms, visible as arrays of darker pixels due to masking. London Array is approximately in the middle of each image. For each image the wind speed and wind direction at 10m for the nearest MERRA2 data point is given for context.

Figures 4 – 9 are provided as examples of the kind of wind field maps possible with this method. External wakes can be seen as darker regions downwind from wind farms in figures 5, 6, 8, and 9. In figure 4 the image is complicated by faster and slower regions of wind speed perpendicular to the wind direction due to gravity waves. In figure 7 gravity waves are the dominant feature to be seen. The wind field is also complicated by wind shadows downwind of land, as seen for example in figures 5 and 7. The complicated spatial nature of the wind speed requires careful consideration of reference wind speeds when quantifying velocity deficits behind wind farms.

2015-12-03 17:40:54

11.0 ms^{-1}, 182°

Figure 4

2016-03-08 17:41:17

7.9 ms^{-1}, 210°

Figure 5

2016-07-03 06:06:19

5.2 ms^{-1}, 258°

Figure 6

IOP Conf. Series: Journal of Physics: Conf. Series **926** (2017) 012004 doi:10.1088/1742-6596/926/1/012004

2016-09-28 17:41:02

9.9 ms^{-1}, 219°

Figure 7

2016-11-18 06:05:41

10.0 ms^{-1}, 247°

Figure 8

2017-03-18 06:05:37

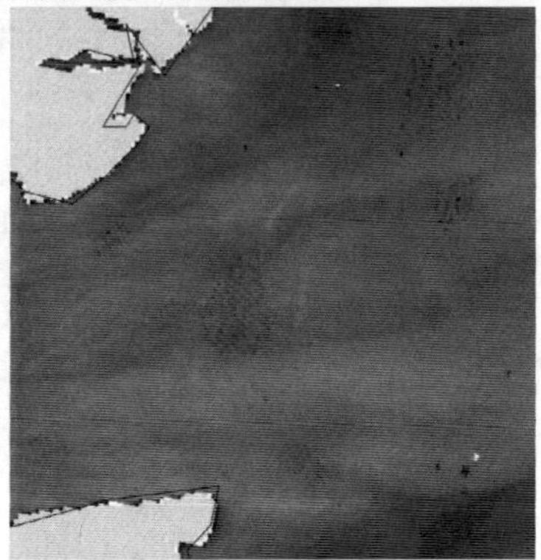

10.5ms^{-1}, 272°

Figure 9

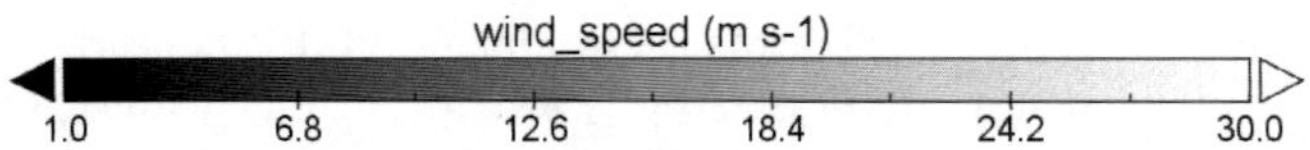

Further work is planned to quantify the effect of variables such as ambient wind speed, coastal proximity and atmospheric conditions on the formation of external wakes at offshore wind farms, looking at many cases over a wide range of locations, seasons and wind conditions in order to separate out as best as possible the effects of the wind farms on the wind flow.

4. Conclusions

This paper sets out a method to derive 500m resolution wind speed maps from Sentinel-1 Synthetic Aperture Radar (SAR) data. The method is tested for accuracy against coincident offshore mast measurements for 25 SAR scenes and a strong correlation is found giving good confidence that the method is robust enough to quantify wind speed differences across offshore wind farms. It does this with free data, tools and software that are available to academia and industry alike.

The process is kept as automatic as possible to facilitate the study of external offshore wind farm wakes over as many scenes as possible. There is much opportunity still to refine and improve the process by trying different wind direction data sources and experimenting with the achievable resolution.

Examples of resulting wind speed maps in the Thames Estuary have been shown. It is anticipated that these wind speed maps will be used to quantify external wake extents and velocity deficits under various conditions.

Acknowledgements

The author of this paper would like to express her thanks to the Centre for Environmental Data Archive for access to the CEDA data archives and the JASMIN computing facility and to the Energy Research Centre of the Netherlands (ECN) for the use of the Meteorological Mast Ijmuiden data. This work was funded by the UK's Engineering and Physical Sciences Research Council (EPSRC) via grant EP/N006224/1 : MAXFARM (MAXimizing wind Farm Aerodynamic Resource via advanced Modelling).

References

[1] European Space Agency 2011 Sentinel 1A: C-band synthetic aperture radar (SAR) data. *Centre for Environmental Data Analysis*. Available : http://catalogue.ceda.ac.uk/uuid/06d1c86f906e42f58172de32c2640be2

[2] Dagestad K-F, Horstmann J, Mouche A, Perrie W, Shen H, Zhang B, Li X, Monaldo F, Pichel W, Lehner S, Badger M, Bay Hasager C B, Furevik B, Foster R C, Falchetti S, Caruso M J and Vachon P 2012 Wind retrieval from synthetic aperture radar - an overview. *Proc. SeaSAR 2012, (Tromso, Norway)*

[3] Zhang B and Perrie W 2012 Cross-polarized synthetic aperture radar, a new potential measurement technique for hurricanes *Bulletin of the American Meteorological Society* **93** 531-41

[4] Zecchetto S, De Biasio F, della Valle A, Quattrocchi G, Cadau E and Cucco A 2016 Wind fields from C- and X-band SAR images at VV polarization in coastal area (Gulf of Oristano, Italy) *Ieee Journal of Selected Topics in Applied Earth Observations and Remote Sensing* **9** 2643-50

[5] Kim T S, Park K A, Li X F, Mouche A A, Chapron B and Lee M J 2017 Observation of wind direction change on the sea surface temperature front using high-resolution full polarimetric SAR data *Ieee Journal of Selected Topics in Applied Earth Observations and Remote Sensing* **10** 2599-607

[6] Hasager C B, Badger M, Mouche A, Stoffelen A, Driesenaar T, Karagali I, Bingöl F, Pena Diaz A, Astrup P, Nielsen M, Hahmann A N, Costa P, Berge E and Bredesen R E 2012 NORSEWInD satellite wind climatology. *DTU Wind Energy E-Report 0007*

[7] Schneemann J, Hieronimus J, Jacobsen S, Lehner S and Kuehn M 2015 Offshore wind farm flow measured by complementary remote sensing techniques: radar satellite TerraSAR-X and lidar windscanners *J. Phys.: Conf. Ser.* **625** 012015

[8] Christiansen M B and Hasager C B 2005 Wake effects of large offshore wind farms identified from satellite SAR *Remote Sensing of Environment* **98** 251-68

[9] Hasager C B, Vincent P, Husson R, Mouche A, Badger M, Pena A, Volker P, Badger J, Di Bella A, Palomares A, Cantero E and Correia P M F 2015 Comparing satellite SAR and wind farm wake models *J. Phys.: Conf. Ser.* **625** 012035

[10] Hasager C B, Vincent P, Badger J, Badger M, Di Bella A, Pena A, Husson R and Volker P J H 2015 Using satellite SAR to characterize the wind flow around offshore wind farms *Energies* **8** 5413-39

[11] Li X M and Lehner S 2013 Observation of TerraSAR-X for studies on offshore wind turbine wake in near and far fields *Ieee Journal of Selected Topics in Applied Earth Observations and Remote Sensing* **6** 1757-68

[12] Charou E, Bratsolis E, Gyftakis S, Giannakopoulos T, Perantonis S and Ieee 2015 Use of Sentinel-1 data for Maritime Domain Awareness : Preliminary results *2015 6th International Conference on Information, Intelligence, Systems and Applications (Iisa)*

[13] Hersbach H, Stoffelen A and de Haan S 2007 An improved C-band scatterometer ocean geophysical model function: CMOD5 *Journal of Geophysical Research-Oceans* **112** C03006

[14] La T V, Khenchaf A, Comblet F, Nahum C and Ieee 2016 Sensitivity of Sea Wind Direction Retrieval from Sentinel-1 Data with Regard to Spatial Resolution and Speckle Noise *2016 Ieee International Geoscience and Remote Sensing Symposium (Igarss)* 3989-92

[15] Gelaro R, McCarty W, Suarez M J, Todling R, Molod A, Takacs L, Randles C A, Darmenov A, Bosilovich M G, Reichle R, Wargan K, Coy L, Cullather R, Draper C, Akella S, Buchard V, Conaty A, da Silva A M, Gu W, Kim G K, Koster R, Lucchesi R, Merkova D, Nielsen J E, Partyka G, Pawson S, Putman W, Rienecker M, Schubert S D, Sienkiewicz M and Zhao B 2017 The Modern-Era Retrospective Analysis for Research and Applications, Version 2 (MERRA-2) *Journal of Climate* **30** 5419-54

IOP Conf. Series: Journal of Physics: Conf. Series **926** (2017) 012005 doi:10.1088/1742-6596/926/1/012005

Influence of small-scale turbulence on cup anemometer calibrations

M Marraccini[1],K Bak-Kristensen[1],A Horn[2],E Fifield[2] and S O Hansen[2]

[1]Svend Ole Hansen ApS, Copenhagen, Denmark
[2]SOH Wind Engineering LLC, Vermont, USA

E-mail: mma@sohansen.dk

Abstract. The paper presents and discusses the calibration results of cup anemometers under different levels of small-scale turbulence. Small-scale turbulence is known to govern the curvature of shear layers around structures and is not related to the traditional under and over speeding of cup anemometers originating from large-scale turbulence components. The paper has shown that the small-scale turbulence has a significant effect on the calibration results obtained for cup anemometers. At 10 m/s the rotational speed seems to change by approx. 0.5% due to different simulations of the small-scale turbulence. The work which this paper is based on, is part of the TrueWind research project, aiming to increase accuracy of mast top-mounted cup anemometer measurements.

1. Introduction

In 2016 the wind energy sector reached 161 GW of installed capacity in Europe, and would, in a normal wind year, produce 352 TW h of electricity enough to cover 12% of the EUs electricity consumption. A single percent (1%) uncertainty in wind speed estimation due to mal-calibrated cup anemometers results in about 3% uncertainty in the corresponding predicted power production [TW h], due to the cubic law of mechanical energy extracted from the wind by a wind turbine. For the year 2015, a 3% uncertainty in wind power production corresponded on the European scene to about 10.5 TW h. At the present market cost (0.04 Euro per kW h) this 1% uncertainty in wind speed estimation corresponds to more than 400 Millions of Euros.

The calibrations of cup anemometers have been carried out extensively by many wind tunnels starting from the beginning of the 90s. The first calibration deviations were up to approximately 10% of measured wind speed between different wind tunnels. During the last years this band has been narrowed down substantially with deviations of the order of 1%. The uncertainties have been reduced by applying different strategies including the usage of larger wind tunnels and the upgrade of the instrumentation, leading to an increased reliability of the measurements.

A deep and thorough investigation on the characteristics of the flow in the wind tunnel represents, at the moment, the path to follow toward the reduction of the calibration uncertainties. Among the characteristics of the flow in a wind tunnel is the turbulence intensity, which may have an influence on cup anemometer calibrations.

According to Kristensen (1992)[6] a sudden increase of the wind speed gives a torque on the rotor which is numerically larger than a sudden decrease of the same magnitude. Since the rotation rate is not in equilibrium with the full-scale turbulent wind, which constantly has many small and fast changes up and down, the rotor will on average tend to run too fast with respect to the mean wind and thus give a positive bias.

Overspeeding in wind tunnels with lower turbulence intensities has been analyzed by Kristensen, Hansen and Hansen (2012)[2]. A turbulence intensity of 1-2% gives a bias of the order of 0.01% to 0.05% indicating that overspeeding is negligible in wind tunnel calibrations at low turbulence intensities.

However the effects of varying small-scale turbulence on cup anemometer calibrations have never been quantified.

The influence of small-scale turbulence on shear layers reattachment has been investigated by Gartshore (1973)[9] and Melbourne (1979)[8]. Tieleman[7] observed that measurements of mean base-pressure coefficients on square-section prisms have a good correlation with the small-scale turbulence (Fig. 1).

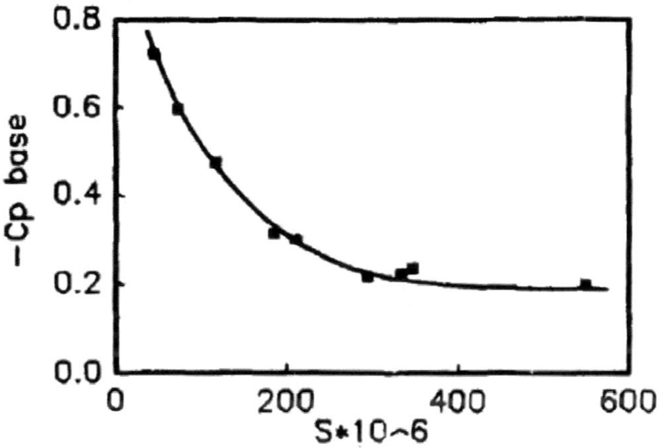

Figure 1: Correlation between mean-base pressure coefficient and small-scale turbulence[7].

Increasing the small-scale turbulence is seen to give numerical smaller suctions on the back side of the structure. For cup anemometer rotors this may affect the forcing both on the driving side and on the resistance side. If the dominating effect of small-scale turbulence occurs on the driving side, the cup anemometer will rotate slower for a specific wind velocity. The cup anemometer will rotate faster if the small-scale turbulence on the resistance side dominates. Thus, whether cup anemometers rotate faster or slower due to small-scale turbulence may depend on the rotor geometry and thereby on the type of cup anemometer considered

2. Calibration uncertainties

The different uncertainty contributions are illustrated by an uncertainty chain in Fig. 2a and Fig. 2b for the year 2000 and 2015, respectively. The chain gives a picture of the particular conditions relevant for cup anemometer calibrations in Measnet approved wind tunnels. The concept of a chain symbolizes that the total uncertainty is governed by the least reliable of the links. The figure also indicates some interactions between different parts of the chain. For both years 2000 and 2015, the uncertainties associated with the measuring time in the wind tunnel and the reading of the cup anemometer frequency are judged to be minor. The uncertainty originating from the instrumentation has been reduced substantially over the last 10-20 years, and this is illustrated in the figure by a distribution being somewhat narrower in 2015 compared to 2000. The same tendency is found for the uncertainty due to air flow quality, and this is due to the benefit of having had Measnet requirements concerning flow quality since 1997, e.g. the Measnet requirements of a small horizontal wind velocity gradient in the wind tunnel. The Measnet requirements have forced the approved calibration wind tunnels to improve their flow quality over the years.

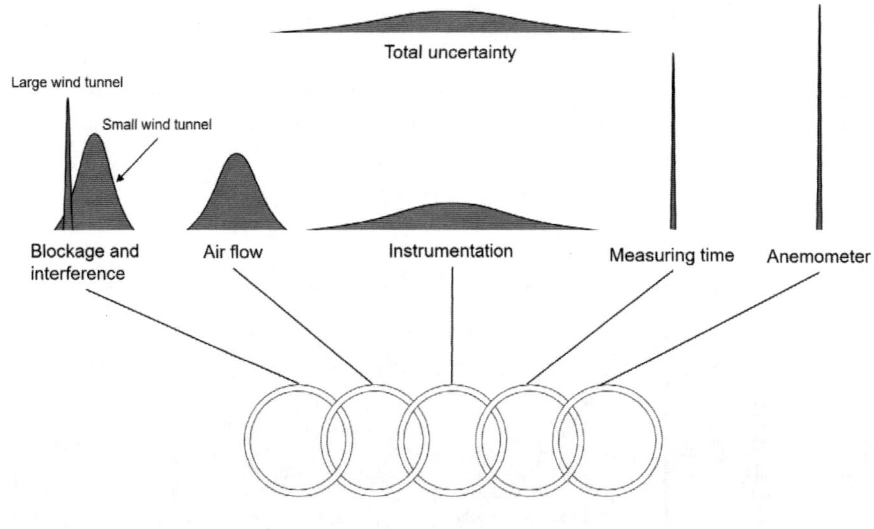

(a)

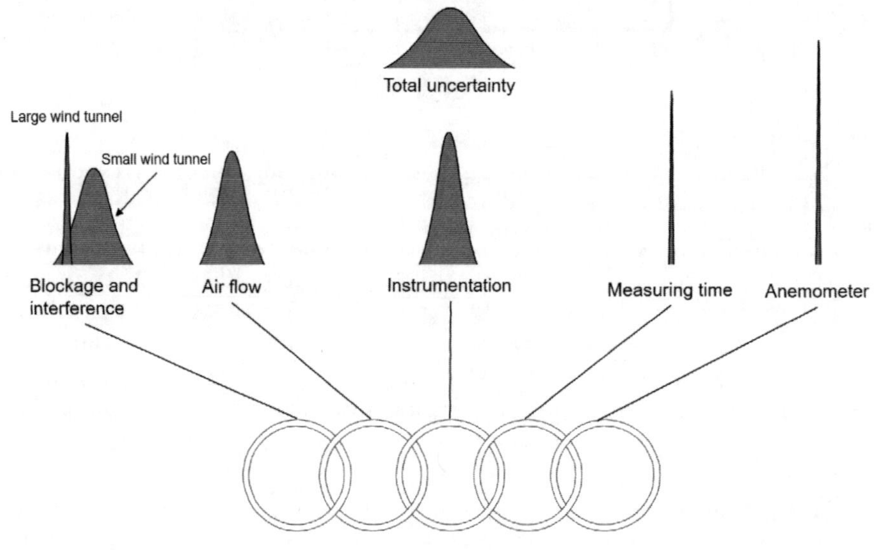

(b)

Figure 2: Illustrations of different uncertainty contributions in 2000(a) and in 2015(b).

The uncertainties originating from wind tunnel blockage and tunnel interference are large for small wind tunnels. However, the opening of two Measnet approved very large wind tunnels in Burlington, Vermont, USA, in 2015 has enabled that the uncertainties originating from wind tunnel blockage and tunnel interference to become negligible in these two wind tunnels. For the two large wind tunnels in Burlington the uncertainties originating from the air flow is not hidden by large uncertainties due to wind tunnel blockage and tunnel interference, and this makes it possible to determine the effect of different air flows on the calibration results without any significant bias from uncertainties originating from wind tunnel size . Thus, the two large wind tunnels in Burlington enable an accurate determination of the effect of small-scale turbulence, and this would have been difficult in smaller wind tunnels because of the dominating uncertainties from other sources.

In spite of good repeatability for a single wind tunnel, different wind tunnels may still have wind speed differences of up to typically 1% between their calibration results obtained for the same cup anemometer, and the differences may depend on the cup anemometer type calibrated. A possible outcome of this is illustrated in Fig. 3.

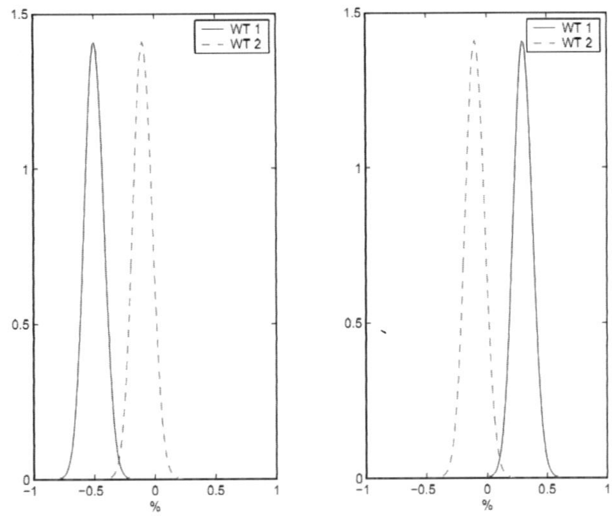

Figure 3: Bias between wind tunnels Njord 1 (WT1) and Njord 2 (WT2) described by distributions.

The main reason for the systematic bias between wind tunnels may be the influence of blockage and the interference between rotating cup anemometer rotors and tunnel boundaries. The typical Measnet wind tunnel is approx. 1 x 1 m in dimension, and this has turned out to be too small for avoiding blockage and tunnel interference effects that may give biased calibrations using the present procedures. There is a profound need of reference calibrations of different cup anemometers carried out in a larger wind tunnel with cross sectional areas of approx. $5\,\text{m}^2$ to $10\,\text{m}^2$, where the effect of blockage and tunnel interference are negligible. A comparison between calibrations in smaller and larger wind tunnels may be used to estimate correction factors for the smaller wind tunnels taking into account their blockage and tunnel interference effects thereby reducing the accompanying calibration uncertainties.

It is highly recommended that Measnet initiates a project focusing on these comparisons. The main outcome expected will be reduced uncertainties of calibration results for cup anemometers, and this follows in a natural way the development of reduced uncertainties obtained by Measnet over the last approx. 20 years, see the discussion above.

3. Investigations carried out

3.1. Wind tunnel setup

SOH Wind Engineering LLC, Vermont, USA, owns two large wind tunnels (Njord 1 and Njord 2) with a test section of 2.5 m × 2.5 m (see Fig. 4). A large amount of cup anemometers of different models and brands are calibrated daily fulfilling IEC[1]61400-12-1[1] whose present guidelines does not include the influence of small-scale turbulence. The turbulence intensity of the flow can be manipulated by applying up to two metal screens in front of the contraction. All the experiments carried out for the present work refer to six different turbulence configurations. The two tunnels have been accredited by Measnet[2] in January 2015 and by IECRE[3] in April 2017. The approved configuration by Measnet and IEC is without screens. The turbulence requirements are however satisfied for the other screen as well giving a turbulence intensity below 2% which including longitudinal wind speed fluctuations with frequencies of up to 5 Hz .

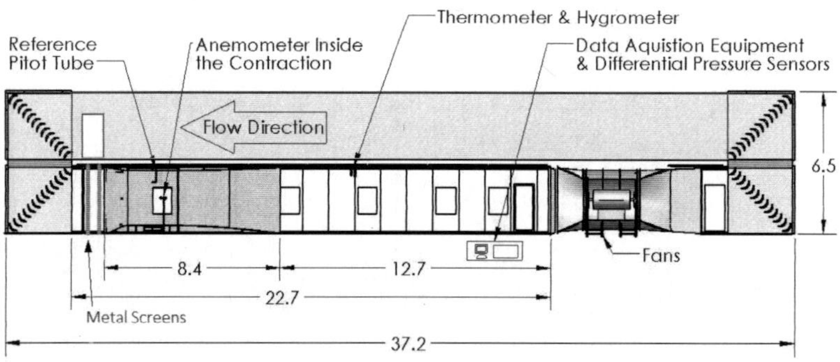

Figure 4: Elevation of wind tunnel, SOH Wind Engineering LLC

Different commercial cup anemometers of various models have been calibrated at SOH Wind Engineering LLC under different turbulence conditions. The characteristics of the flow in the calibration point of both tunnels have been measured for different wind velocities through hot-wire anemometers able to measure wind speed fluctuations with a sampling rate up to 10 kHz. Each time series has been used to calculate the power spectral density of the longitudinal velocity component up to 5 kHz according to Nyquist-Shannon theorem. The investigated wind speeds span a range from 4 m/s to 16 m/s, this range being applied for anemometer calibrations according to IEC 61400-12-1.

3.2. Small-scale spectral density parameter

Introduced by Melbourne (1975), the small-scale spectral density parameter S is defined as the spectral density of the longitudinal velocity $nS_u(n)$ (n is the frequency in hertz) at a wavelength U/n corresponding to one tenth of the characteristic length scale $\sqrt{A}$ of the model, normalized by the squared mean wind velocity U^2 (Eq. 1).

$$S = \frac{nS_u(n)}{U^2} \qquad \frac{U}{n} = 0.1\sqrt{A} \tag{1}$$

Fig. 5 shows the small-scale spectral density parameter at 10 m/s for different configurations. The characteristic length scale $\sqrt{A}$ has been assumed to be 0.13 m. The vertical black line indicates the calculated frequency at 10 m/s.

[1] International Electrotechnical Commission
[2] International network for harmonised and recognized measurements in wind energy
[3] IEC system for certification to standards relating to equipment for use in renewable energy applications

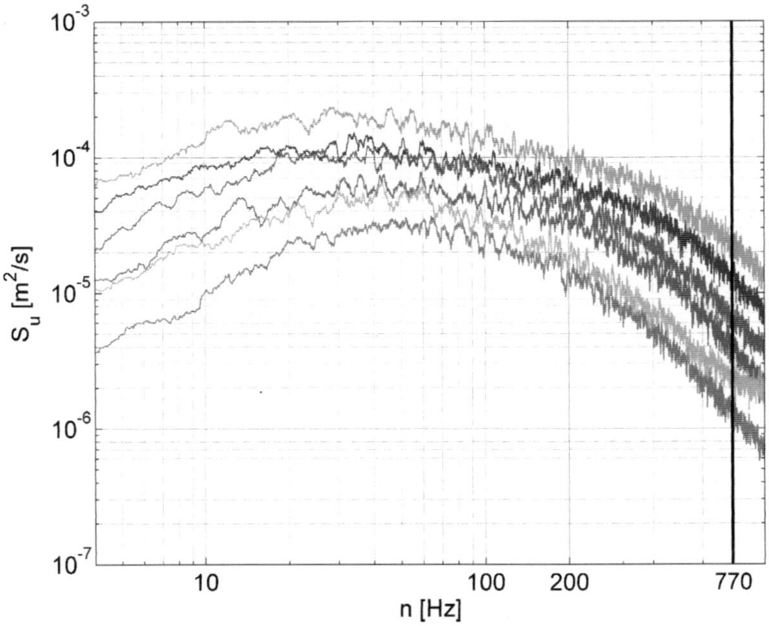

Figure 5: Small-scale spectral density parameter for 10 m/s and different configurations. Njord 1 Two screens (red), Njord 2 Two screens (yellow), Njord 1 One screen (green), Njord 2 One Screen (violet), Njord 1 No screens (blue), Njord 2 No screens (cyan).

3.3. Anemometer calibration

The calibration expression of a cup anemometer can be rearranged as Eq. 2

$$f(v_t) = \frac{v_t - k_b}{k_a} \tag{2}$$

where $f(v_t)$ is the output frequency of the anemometer at the target wind speed v_t, k_a is the slope and k_b is the offset. The comparison is based on the frequency which an anemometer would indicate at a target wind speed v_t.

4. Results

A target wind speed of 10 m/s has been defined and the ideal frequency of each anemometer for each turbulence configuration has been calculated. In order to isolate the dependency on the small-scale turbulence, each anemometer has been analysed independently. This can be achieved by normalising the frequencies corresponding to different experiments (turbulence configurations and different tunnels) by the mean value of all experiments performed on a specific anemometer. This is done in order to investigate the global trend of all the normalized frequencies with respect to the small-scale turbulence as different brands/models of anemometers have different calibration expressions.

The small-scale turbulence parameter can be calculated for a wind speed of 10 m/s based on power spectral density results according to Eq. 1. Fig. 6 shows six vertical bars corresponding to six levels of small-scale turbulence. The normalized frequencies of all anemometers are always contained in the vertical bars. Points where the normalized frequencies are above 1 show that the rotor of the anemometer rotates faster than the average value and vice-versa.

Fig. 6 shows that the investigated cup anemometers rotate slower when the small-scale turbulence is increased. This tendency is found both for cup anemometers rotating clock-wise and anti-clockwise indicating that the change of rotational speed is not caused by non-desired horizontal gradients of the air flow introduced when the screens are installed in the wind tunnels.

The results in Fig. 6 depict a global trend of the averaged frequencies which decrease as the small-scale parameter increases. The S parameter magnitudes shown in Fig. 1 and Fig. 6

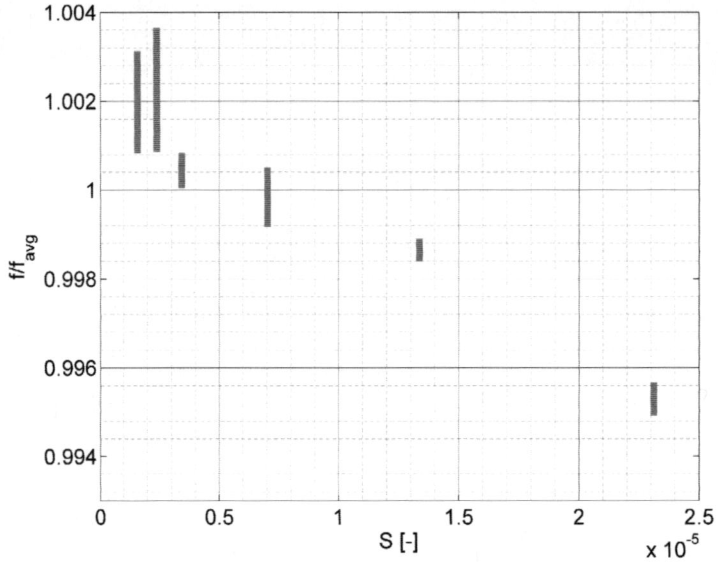

Figure 6: Normalized frequencies as function of the small-scale turbulence S at a target wind speed of 10 m/s

cannot be directly compared as they are calculated for different bodies and under different wind conditions. However the overall trend of the two figures can be compared showing a good correlation between the experiments, assuming that small-scale turbulence dominates the driving side of the rotor.

The absolute deviation of the normalized frequencies between the different turbulence configurations is of the order of 0.5% - 1% whereas small variation of turbulence intensity have been seen to give a bias of the order of 0.01% to 0.05% [2]. The results shown in Fig. 6 cannot be attributed to variations of turbulence intensity indicating that the small-scale turbulence becomes responsible for non-negligible deviations of calibration results.

The wide bars at low levels of small-scale turbulence may originate from different dependencies for different different anemometer brands/models. This evidence can be compared with the results obtained by Tieleman in Fig. 1: lower values of small-scale turbulence correspond to the most inclined sector of the experimental curve, i.e. where small variations of small-scale turbulence can give large differences of mean base-pressure. This is consistent with the observations of the present paper.

5. Conclusions

The paper has shown that the small-scale turbulence has a significant effect on the calibration results obtained for cup anemometers. At 10 m/s the rotational speed seems to change by approx. 0.5% due to different simulations of the small scale turbulence. It is emphasized that the effect of small-scale turbulence originates from changed forcing acting on the rotor, and it is not possible to take this effect into account by the present mathematical models of cup anemometer rotors, e.g. the models applied in cup anemometer IEC classifications [1].

In the short term it is recommended that the Measnet approved calibration wind tunnels quantify their small-scale turbulence, and this information may in the long term lead to requirements for small-scale turbulence in Measnet and subsequently IEC specifications. These requirements may follow Fig. 7b , where full-scale could be a typical full-scale turbulence condition with normal turbulence. On long term manufactures of cup anemometers may use a small sensitivity to small-scale turbulence as a competitive advantage.

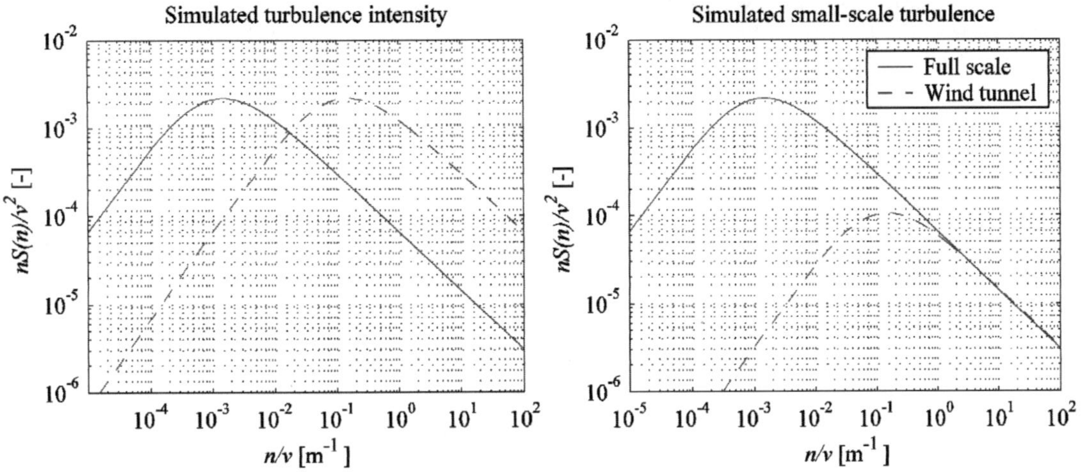

(a) (b)

Figure 7: (a) shows a wind tunnel simulation based on a correct representation of the turbulence intensity assumed to be 10%. (b) shows a wind tunnel simulation with reduced turbulence intensity, but a correct representation of the small-scale turbulence. The turbulence intensity in (b) is approx. 2%.

A thorough understanding of all aspects influencing calibrations of cup anemometers is the first step towards increasing the accuracy of wind speed measurements in field. The work presented in the paper regarding the small-scale turbulence influence on the calibrations is a breakthrough in regards to using wind tunnel calibrations results to interpret wind speed field measurements as well as reducing the bias between calibration results obtained in the individual anemometer calibration institutes. This is of great importance in order to meet the demands from the wind energy industry on a reduction of uncertainty regarding the wind speed measurements.

6. Learning objectives

The outcome of the presented work is firstly of great relevance for the anemometer calibration institutes, creating an awareness of the influence of turbulence on the calibration results. The turbulence influence is an aspect well known in other parts of wind tunnel testing experiments but is yet to be recognized within the cup anemometer calibration industry. The present guidelines for cup anemometer calibrations refer alone to the turbulence intensity of the flow not handling the influence of small-scale turbulence and this contributes to the present well known bias between the individual anemometer calibration institutes. The paper aims to spread the achieved knowledge of small-scale turbulence influence for the general benefit of reduction of uncertainties on cup anemometer calibrations.

7. Acknowledgement

The presented work is part of the TrueWind research project, funded by EUDP (*Det Energiteknologiske Udviklings og Demonstrationsprogram*: a founding program under the Danish Energy Agency).

References

[1] *International Standard 61400-12-1, 2017, Second Edition* International Electrotechnical Comission (IEC), Geneva, Switzerland.

[2] Kristensen L, Hansen O F and Hansen S O *Addendum to: Bias on horizontal meanwind speed and variance caused by turbulence* www.windsensor.dk, May 29, 2012

[3] Kristensen L and Hansen O F *Bias on horizontal mean-wind speed and variance caused by turbulence* www.windsensor.dk, May 27, 2012

[4] Hansen O F, Hansen S O and Kristensen L. *Wind tunnel calibration of cup anemometers* AWEA Wind Power, Atlanta, 2012.

[5] Dyrbye C and Hansen S O *Wind loads on structures* Wiley,1997.

[6] Kristensen L *The cup anemometer and other exciting instruments* Ris National Laboratory, Roskilde, Denmark, April 1992.

[7] Tieleman H W *Problems associated with flow modelling procedures for low-rise buildings* Journal of Wind Engineering and Industrial Aerodynamics, 1992.

[8] Melbourne W H *Turbulence Effects on Maximum Surface Pressures - A Mechanism and Possibility for Reduction* Ed. J. E. Cermak, Vol. i, pp. 541-551, 1979, Pergamon Press, New York, N.Y., 1979.

[9] Gartshore I S *The effects of freestream turbulence on the drag of rectangular two dimensional prism* University of Western Ontario, London, Ontario, Canada, 1973.

Improving uncertainty estimates: Inter-annual variability in Ireland

D Pullinger, M Zhang, N Hill and T Crutchley

Lloyd's Register EMEA, 27 Gay Street, Bath BA1 2PD, United Kingdom

david.pullinger@lr.org

Abstract. This paper addresses the uncertainty associated with inter-annual variability used within wind resource assessments for Ireland in order to more accurately represent the uncertainties within wind resource and energy yield assessments. The study was undertaken using a total of 16 ground stations (Met Eireann) and corresponding reanalysis datasets to provide an update to previous work on this topic undertaken nearly 20 years ago. The results of the work demonstrate that the previously reported 5.4% of wind speed inter-annual variability is considered to be appropriate, guidance is given on how to provide a robust assessment of IAV using available sources of data including ground stations, MERRA-2 and ERA-Interim.

1. Introduction

1.1 Wind resource assessment

In order to calculate the financial value of a proposed wind energy development, before construction takes place, a robust understanding of the site's wind resource is required. The revenue of a project is derived from the energy that the project can produce and the price at which that energy can be sold.

The energy produced by wind turbines comes from the wind, hence the requirement for a wind resource assessment. This field which relies on principles of meteorology, physics, statistics and engineering is one that is developing rapidly and has been over the last 30 years. The typical wind resource assessment includes the following steps:

- On-site measured data analysis;
- Long-term wind resource assessment;
- Wind flow modelling;
- Loss assessment;
- Uncertainty assessment.

On-site measurements of the wind parameters of the site, are typically only available for a period of one to two years, this is a limited period of time when compared to the likely 20-year operational lifetime of a wind energy project. Wind speed varies from year to year and therefore a long-term adjustment of the on-site measured data is desirable.

In addition to the long-term adjustment the inter-annual variability needs to be understood in order for the uncertainty related to these predictions to be quantified. Greater understanding of the long-term wind resource and the inter-annual variability will lead to more robust predictions of the financial value of a wind energy project.

1.2 Inter-annual variability
Inter-annual variability (IAV) is a measure of the extent to which wind speeds vary from one year to the next. IAV is defined as the standard deviation of annual mean wind speed over several years. This value is typically represented as a percentage. Inter-annual variability typically impacts on "anywhere between 10 % and 25 % (figures representing both historic and future wind variability) of the overall uncertainty" [1] over a 10-year return period.

Whilst the calculation of IAV is simple, deriving a site-specific or even regional representative value is not so straightforward. Wind measurements for development projects are typically carried out on-site over a short term period using a meteorological mast, and more recently remote sensing devices such as SoDAR (Sonic Detection and Ranging) and LiDAR (Light Detection and Ranging) devices. These campaigns last a minimum of 12 months, and after this point it is usually in the interest of the project developer to achieve construction as soon as possible. For this reason the majority of site measurement campaigns are of short duration and therefore a site specific figure of IAV using on-site measurements alone is not possible. If there are no on-site long-term sources, then an alternative source of long-term reference data is required.

1.3 Long-term reference sources
There are two primary types of wind data source used for long-term correction of measured data. Firstly: actual measurements, usually taken by a national meteorological organisation using anemometers. Secondly: virtual datasets, consisting of data from models (both reanalysis and mesoscale models).

With any source of long-term reference there are two key criteria for their use within a long-term assessment. Firstly they need to be consistent – providing a true representation of the long-term variation historically. Secondly they need to be representative of the site being modelled – often measured through criteria such as correlation.

1.3.1 Actual measurements In Ireland the primary source of long-term reference data comes from Met Eireann, Ireland's National Meteorological Services. Met Eireann has a network of 25 surface meteorological stations spread across the country; these typically consist of a meteorological mast with an anemometer mounted on it at a height of 10 m above ground level.

Each meteorological station has metadata available (logs detailing sensor changes, mast movements and exposure changes). Using the metadata it is possible to make a judgement to ensure that the readings from the station can be considered consistent.

1.3.2 Virtual datasets Reanalysis datasets take historical meteorological measurements from a variety of sources, including surface-based stations, rawinsonde, satellites, aircraft measurements, etc. and apply a model which is used to predict the states of the weather at different times and locations. One output of some of these models is the surface wind speed – which is of use for wind energy professionals as a source of long-term reference.

Within this study two commonly used reanalysis models are considered: 1) The MERRA-2 (Modern Era Reanalysis for Research and Applications-2) [2] datasets with a horizontal resolution of 0.5° x 2/3°

and a temporal resolution of an hour; 2) The EMD-ERA (based on ERA-Interim) datasets [3] with a horizontal resolution of approximately 80 km and a temporal resolution of 6 hours.

One of primary limitations of reanalysis data sources in relation to assessing long-term wind speeds for a site is whether the model can be considered to be consistent. Several papers have been published attempting to validate the consistency of reanalysis data sources over time [4], [5], [6], [7]. The use of reanalysis data sources has been previously studied in South Africa [1] and this study applies a similar approach to Ireland.

1.4 Previous Studies

The literature available on the inter-annual variability of wind speeds internationally is fairly limited. From experience the current industry standard is to use a figure of ~6 % (annual mean wind speed) that is applicable to North Western Europe and based on a study from 1997 [8]. This study presents an IAV figure for Ireland of 5.4 % [8], based on five meteorological stations covering a cumulative time period of 144 years.

Studies of the IAV in the United States [9] and South Africa [1] have recently been published. Both of these studies examine whether reanalysis sources may be appropriate for representing the IAV in these two regions. The South Africa study concluded that whilst reanalysis models may be of use in informing the site IAV, adjustments are required as there may, in some cases, be under prediction of the IAV.

1.5 Summary of Objectives

The study aims are to address the following areas:

- Determine a general IAV figure (or range) applicable to Ireland – in order to compare this to the current industry standard of 5.4%. The results may challenge historic industry assumptions.
- Determine if a consistent relationship exists between the IAV derived from ground stations and the reanalysis models, informing of the ability of global reanalysis models to guide the site IAV assumptions.
- Determine the length of data period required for the IAV to converge on a consistent value.

The aim of the study is to provide practical guidance on the best approach to evaluate IAV in Ireland for wind energy projects. The results will be useful for stakeholders within the wind energy industry including wind resource analysts, developers, owners, operators, meteorologist, research institutes and financial investors.

2. Methodology

2.1 Input Data

Long-term wind data from all 25 Met Eireann stations has been obtained. For each of the Met stations, the nearest MERRA-2 [2] and ERA-Interim [3] nodes have also been downloaded. For these nodes a height above ground level of 50 m was used throughout the analysis.

2.2 Data Screening

Following the accumulation of input data a screening process was undertaken in order to remove datasets which contained inconsistencies or were not of sufficiently long a period to be of use.

Station inconsistencies were detected through the use of a changepoint analysis method [10] and metadata review. The metadata review consisted of examination of the records for each station in order to identify the following:

- Changes in the measurement equipment, such as anemometer make or model, mast location or data recording system.
- Changes in the exposure surrounding the site, such as growing trees, new buildings or obstacles located near to the measurement station.

The changepoint analysis method is documented within [11] and the application of the approach also recorded through [10]. The changepoint analysis represents a statistical technique, based on bootstrapping, which calculates a probability of inconsistency within a time series of data, this technique when applied to monthly-mean wind data has been demonstrated to have a high accuracy, and particularly useful for wind analysis a low false negative rate [10].

Following the screening process 16 datasets remained for use within the study. These are displayed in Figure 1.

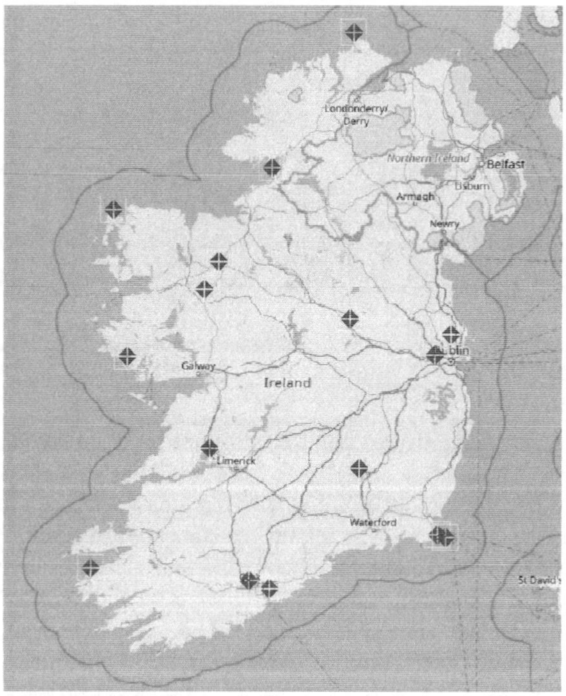

Figure 1. Locations of reference stations studied
© OpenStreetMap CC-BY-SA

2.3 IAV Calculation

Following the completion of the site identification and screening processes the IAV was calculated for each site from the ground stations and nearby MERRA-2 and ERA-Interim nodes. The IAV was calculated both for the concurrent period with the ground station and also for the full period. This allows for an understanding of whether the period of ground measurements could be considered as representative of the long-term.

2.4 Relationship between datasets
The second objective of the work was to determine whether any relationship between the datasets existed, in terms of their IAV. A comparison between the short-term and long-term IAV figures was calculated for each site, this was performed as a simple ratio of short-term to long-term.

2.5 Period required for convergent IAV
In order to determine the period of data required for the IAV figure to converge, a cumulative (going backwards in time) standard deviation was calculated. This started with the most recent three whole years (2014 to 2016) and was extended by including additional years until the length of the reference dataset was reached.

The IAV development over period was plotted and convergence was considered to be achieved at the period in which the IAV value is within +/- 15 % of the full period IAV, this approach is consistent with that from a previous study of South Africa [1]. Within this approach there is a trade-off between reaching a fully converged figure and the number of data sources with the required length of data available. Studies have shown the large drop off in the number of available consistent reference station sources as time period increases [7]. For obvious reasons, such as improving technology and greater understanding of the importance of consistency, wind speed measurements going back further than 20-years are often not of a similar quality to those from modern sensors. For reanalysis sources this is not typically a problem (as the data sources consist of data covering a long historic period of usually greater than 30 years), however for comparison with ground stations a higher convergence figure is required.

3. Results

3.1 IAV calculation
The results are presented in Table 1. following the aforementioned screening of the datasets, these results were based on a total of 174 years of measured wind data. It should be noted that these results are a percentage of wind speed.

Table 1. Minimum, Mean and Maximum IAV for each studied data source

	Ground Station	MERRA-2	ERA-Interim
Minimum	4.4 %	3.0 %	3.0 %
Mean	5.4 %	4.1 %	4.0 %
Maximum	6.9 %	5.7 %	4.7 %

3.2 Relationship between datasets
The results demonstrate that across all sites MERRA-2 and ERA-Interim under-predict the IAV at the corresponding ground station. The results vary from approximately 50 % to 100 % of the ground station IAV with both of reanalysis sources producing very similar mean figures – these can be seen in Table 2.

Table 2. Minimum, Mean and Maximum ratios between reanalysis and ground stations

	MERRA-2: Ground Station	ERA-Interim: Ground Station
Minimum	58.2 %	56.7 %
Mean	76.8 %	74.3 %
Maximum	100.0 %	89.6 %

3.3 Period required for convergent IAV

Following the methodology outlined previously, the change in IAV over time was calculated for each site and reanalysis source. An example is given in Figure 2, the overall results for all 16 sites are given in Table 3. This plot shows how the IAV converges on the long-term figure and +/- 15% bounds are applied, the 15 % convergence is selected based on previous studies [1]. It can be seen that for this site convergence is achieved after 11 years.

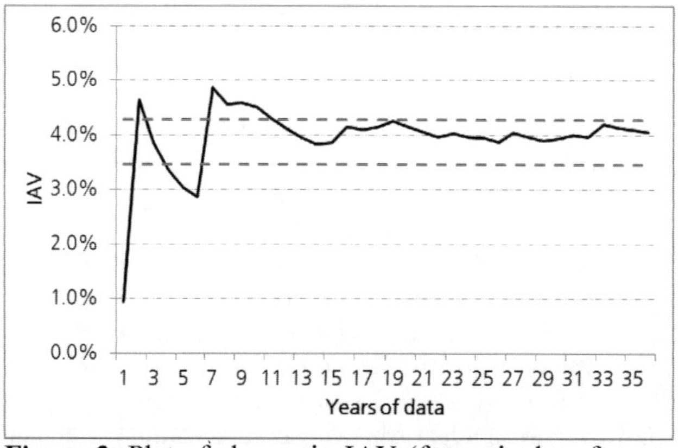

Figure 2. Plot of change in IAV (for a single reference site) with increasing data period (black line), +/- 15 % limit for convergence plotted (red dashed line)

Table 3. Data period (years) required for converged IAV

	Ground station	MERRA-2	ERA-Interim
Minimum	4.0	6.0	7.0
Mean	8.8	8.5	10.4
Maximum	13.0	11.0	23.0

4. Conclusions

4.1 IAV in Ireland

The mean IAV across the Met Eireann ground stations was calculated to be 5.4 %. This agrees with the previously calculated figure from a study in 1998 [8]. The previous study, whilst considering longer time periods from each of the five stations used, was limited in the spatial coverage. The combination of these two studies – greatly increases confidence in the IAV figure. The agreement between the two studies is re-assuring, whilst some of the ground stations considered within are the same (although covering different time periods) some stations are different between the studies. Combined the two studies include a total of 318 years of reference data from 16 independent sources. This is considered a comprehensive sample to represent the IAV.

It is noted that the IAV figure does vary between sites with a range from 4.4 % to 6.9 % (for reference the previous study [8] ranged from 4.7 % to 6.4 % across the five ground stations). From analysis it is not clear as to the cause of this variation, no clear trends are observed in terms of coastal against non-coastal or geographically. Local measurements are recommended wherever possible, however it is considered that the mean figure for Ireland of 5.4 % is a robust assessment and can be used within energy yield assessments.

4.2 Reanalysis data as a source of IAV

The study has examined how well the reanalysis data sources MERRA-2 and ERA-Interim represent the IAV measured at ground stations. A strong correlation is not measured ($R^2 < 0.5$) between either source and the surface stations. However from the results it is concluded that both reanalysis models under-predict the IAV, typically by 25 %. Care should therefore be taken when considering reanalysis sources within any assessment of site specific IAV.

The lack of relationship between reanalysis data and ground measurements agrees with a study in South Africa [1]. It is speculated that the reanalysis inter-annual variability is lower than this derived from measurements due to the setup of the model, that typically reanalysis does not pick up on inter-decadal variation particularly well and therefore this results in a lower result.

4.3 Period required for convergent IAV

From this study it is recommended that a data period of a minimum of 10 years is used in order to derive the IAV when using any source of data. This agrees, and is slightly less conservative than, a similar study performed for South Africa [1]. It is considered that the +/-15 % figure representing convergence is appropriate due to the impact this level of error is likely to have on the overall wind resource assessment uncertainty. The impact is not negligible; however the authors consider this a reasonable compromise between accuracy and number of available data sources.

4.4 Practical guidance for wind resource assessment

The purpose of this study has been to inform more robust wind resource and energy yield assessments, therefore a few recommendations are given below which should be considered within these analyses.

- Screening of reference sources for consistency is important to ensure representative results.
- When investigating IAV a period of >10 years should be considered to provide a representative sample.
- A figure of 5.4 % IAV in Ireland is considered appropriate, however this figure may be refined if local long-term site data is available.
- MERRA-2 and ERA-Interim under-predict IAV, however the level of under-prediction varies and this should be considered if using these data sources to predict IAV.
- A comprehensive justification of the IAV figure should be provided within any wind resource or energy yield assessment in order to allow for thorough review and due diligence.

4.5 Further work

It is becoming clear from the four main studies into the area of inter-annual variability, covered in this study, the USA [9], South Africa [1] and Offshore UK [12], that reanalysis models behave differently across the globe. Therefore a simple global study is considered unlikely to be of sufficient depth to capture local effects. Therefore it is recommended that for new regions this study is repeated in order to ensure that IAV assumptions are accurate within uncertainty assessments.

Addition of further sources of wind data, including operational wind farm data would add benefit to this study. This could confirm the representativeness of the surface stations as a measure of wind farm production variability due to wind speed for a specific wind farm development.

5. References

[1] Behrens M and Pullinger D 2016 Inter-annual variability of wind speed in South Africa *Wind Prospect*

[2] Rienecker M, Suarez, M, Gelaro G, Todling R, Bacmeister J, Liu E, Bosilovich M, Schubert S,

Takacs L, Kim G *et al.* 2011 NASA's Modern-Era Retrospective Analysis for Research and Applications *J. Climate* **24** pp. 3624-3648

[3] Dee D, Uppala S, Simmons A, Berrisford P, Poli P, Kobayashi S, Andrae U, Balmaseda M, Balsamo G, Bauer P *et al.*, 2011 The ERA-Interim reanalysis: configuration and performance of the data assimilation system *Quarterly Journal of the Royal Meteorological Society* **137:656** pp. 553-597

[4] Brower M 2006 The use of NCEP/NCAR reanalysis data in MCP *Procs. EWEC 2006* (Athens: EWEC)

[5] Jiminez B, Moennich K and Durante F 2012 Comparison of NCEP/NCAR and MERRA reanalysis data for long-term correction in wind energy assessment *Procs. EWEA 2012* (Copenhagen: EWEA)

[6] Lileo S and Petrik O 2011 Investigation of the use of NCEP/NCAR, MERRA and NCEP/CFSR reanalysis data in wnd resource analysis *Procs. EWEA 2011* (Brussels: EWEA)

[7] Pullinger D and Davies O 2012 Validation of MERRA data as long-term reference source in Great Britain *Procs. RenewableUK 2012* (Glasgow: Renewable UK)

[8] Raftery P, Tindal A and Garrad A 1997 Understanding the risks of financing windfarms *Procs. EWEA Wind Energy Conference* (Dublin: EWEA)

[9] Brower M, Lledo L, Barton M and Dubois J 2013 A study of wind speed variability using global reanalysis data *AWS Truepower*

[10] Pullinger D, Zhang M, Hill N and Crutchley T 2017 Novel statistical validation of long-term consistency of reanalysis models in South Africa *Procs. WIndAc 2017* (Cape Town: WindAc)

[11] Cowan A 2017 A review of wind speed inter-annual variability for the UK offshore wind climate *Procs. Global Offshore Wind* (London: WindEurope)

[12] Candlish G 2012 Interannual wind variation from observations and numerical weather analyses *UKMO*

[13] Pullinger D and Hill N 2015 Variability of variability, how well does MERRA and ERA-Interim map inter-annual variability *Procs. EWEA Resource Assessment Forum* (Helsinki: EWEA)

Acknowledgements

All MERRA, ERA-Interim and ground station datasets were downloaded using WindPRO software v3.0.629 developed by EMD International A/S: http://www.emd.dkor http://www.WindPRO.com.

Lastly, the authors would like to acknowledge the Global Modelling and Assimilation Office (GMAO) and the GES DISC (Goddard Earth Sciences Data and Information Services Center), as well as the European Center for Medium-Range Weather Forecasts for the dissemination of MERRA and ERA-Interim.

Multi-step-ahead Method for Wind Speed Prediction Correction Based on Numerical Weather Prediction and Historical Measurement Data

Han Wang[1], Jie Yan[1*], Yongqian Liu[1], Shuang Han[1], Li Li[1], Jing Zhao[2]

School of Renewable Energy, North China Electric Power University, Beijing, 102206, China [1]

Beijing Shenzhou Aerospace Software Technology Co., Ltd., Beijing, 100094, China [2]

E-mail: yanjie_freda@163.com

Abstract. Increasing the accuracy of wind speed prediction lays solid foundation to the reliability of wind power forecasting. Most traditional correction methods for wind speed prediction establish the mapping relationship between wind speed of the numerical weather prediction (NWP) and the historical measurement data (HMD) at the corresponding time slot, which is free of time-dependent impacts of wind speed time series. In this paper, a multi-step-ahead wind speed prediction correction method is proposed with consideration of the passing effects from wind speed at the previous time slot. To this end, the proposed method employs both NWP and HMD as model inputs and the training labels. First, the probabilistic analysis of the NWP deviation for different wind speed bins is calculated to illustrate the inadequacy of the traditional time-independent mapping strategy. Then, support vector machine (SVM) is utilized as example to implement the proposed mapping strategy and to establish the correction model for all the wind speed bins. One Chinese wind farm in northern part of China is taken as example to validate the proposed method. Three benchmark methods of wind speed prediction are used to compare the performance. The results show that the proposed model has the best performance under different time horizons.

Key words. Wind speed prediction, multi-step-ahead, prediction correction, time-dependent effects, support vector machine.

1. Introduction

As an important renewable energy resource, wind power has been increasing dramatically in the power system worldwide [1]. However, the intermittence of wind brings serious challenges to the security and stability of the power system operation as well as the power quality [2-3]. Cubic relationship between wind speed and power output means that even tiny wind speed prediction error would trigger very large wind power forecasting deviation. To improve the wind speed prediction accuracy on both short-term and ultra-short-term is one of the effective ways to reduce the negative impact of wind power integration on the power system [4]

Previous literatures on wind speed prediction can be divided into two categories [5]. The first category is based on historical wind speed measurements to make ultra-short-term prediction, primarily 1-4h. Its main idea is to mapping the internal relationship between wind speed time series.

The commonly used algorithms include ARMA [6-7], Kalman filter [8], gray theory [9], etc. However, the accuracy of such methods fall sharply with the increase of prediction time horizon since the impact of previous wind series on the following wind condition is weakened. Another category employs the Weather Research and Forecasting (WRF) mesoscale NWP wind speed [10]. The initial and the boundary field of the WRF model is released by the National Centers for Environmental Prediction at 6 o'clock every day. The WRF model is adopted to extract the time-serial results from 24 o'clock a day to 24 o'clock the next day as the input data for the power forecasting. The accuracy of NWP data suffers relatively small impacts from the prediction horizons, however, the deviation of NWP is not negligible and contributes largely to the wind power forecasting error. Therefore, the improvement and correction of NWP data in either physical or statistical manner is significant to serve well for the wind power forecasting.

Many researchers are working on the statistical correction of wind speed prediction, and traditionally to establish a mapping relationship between NWP wind speed and the measured wind speed at the same time slot. Commonly used algorithms include neural network [11], SVM [12-13], relevance vector machine [14], and deep learning [15], etc. These methods achieve the accuracy improvement by around 1-2m/s on average, and is slightly affected by the prediction time horizons. However, this mapping relationship, which can be interpreted as NWP deviation pattern, is not always very clear and consequently affects the correcting performance.

To solve above problem, this paper presents a multi-step-ahead wind speed prediction correction method considering the impacts passing from the wind speed in the latest time series. Both numerical weather prediction and historical measurement data are used as training inputs. Additional inputs encourage the NWP deviation pattern to be differentiated under various wind conditions during the training process and improve the correction performance. The results show that the proposed method outperforms the other three benchmark methods at different time horizons, and the accuracy degradation along with the time horizons is reduced to some extent.

The remainder of this paper is organized as follows. Section 2 analyses the probability density of NWP deviation. Section 3 describes the proposed wind speed prediction correction method construction. Section 4 discusses the prediction results among the proposed method and three benchmark methods. Section 5 concludes the paper.

2. Probability Density of NWP Deviation

Numerical weather prediction is the basic inputs of wind power forecasting and is also the main source of the power forecasting error. This section takes one wind farm in north China as example to analyse the probability density of NWP deviation in each wind speed bin and the dependency of previous time series to the following wind speed. The NWP deviation is calculated as eq. (1).

$$v_{deviation}(t) = v_{NWP}(t) - v_m(t) \tag{1}$$

where, $v_{NWP}(t)$ is the NWP wind speed at time of t; $v_m(t)$ is the measured wind speed at time of t.

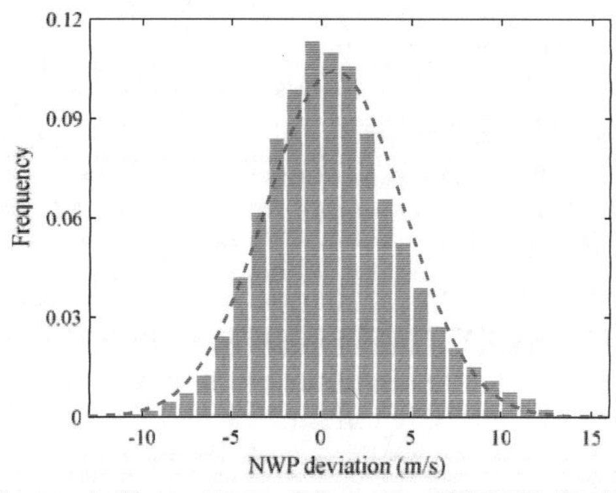

Figure 1. Frequency of the overall NWP deviation

Firstly, the overall NWP deviation in this wind farm is calculated. The maximum value of the positive deviation is 15.8 m/s, the average value of the positive deviation is 3.38m/s. The maximum value of the negative deviation is 12.5 m/s, the average value of the negative deviation is 2.54m/s. The frequency of the overall NWP deviation is shown in Fig. 1. Distribution of the NWP deviation can be fitted by using normal distribution. However, the range of NWP deviation is large, so further study on the deviation of NWP wind speed is needed.

Secondly, the NWP wind speed is segmented into several bins, for example, 1 m/s per bin. According to the range of NWP wind speed and the power curve, four wind speed bins (0-1m/s, 7-8m/s, 14-15m/s, 21-22m/s) are selected to show the results of NWP deviation. Fig.2 shows the frequencies of NWP deviation in several wind speed bins. Tab. 1 shows the statistics of NWP deviation in several wind speed bins, including maximum positive deviation, average positive deviation, maximum negative deviation and average negative deviation. It can be seen from Fig. 2 and Tab. 1, the frequencies of NWP deviation in small bins show different shapes, and cannot be fitted such well by the normal distribution. Also, the range of the NWP deviation is still large even if a small NWP wind speed bin is considered. It means that the relationship between NWP wind speed and the measured wind speed at the same time slot is not very clear. That is one important difficulty for the traditional NWP correction methods to achieve good performance. Therefore, more impact factors need to be considered to be learning labels during the training of wind speed prediction correction.

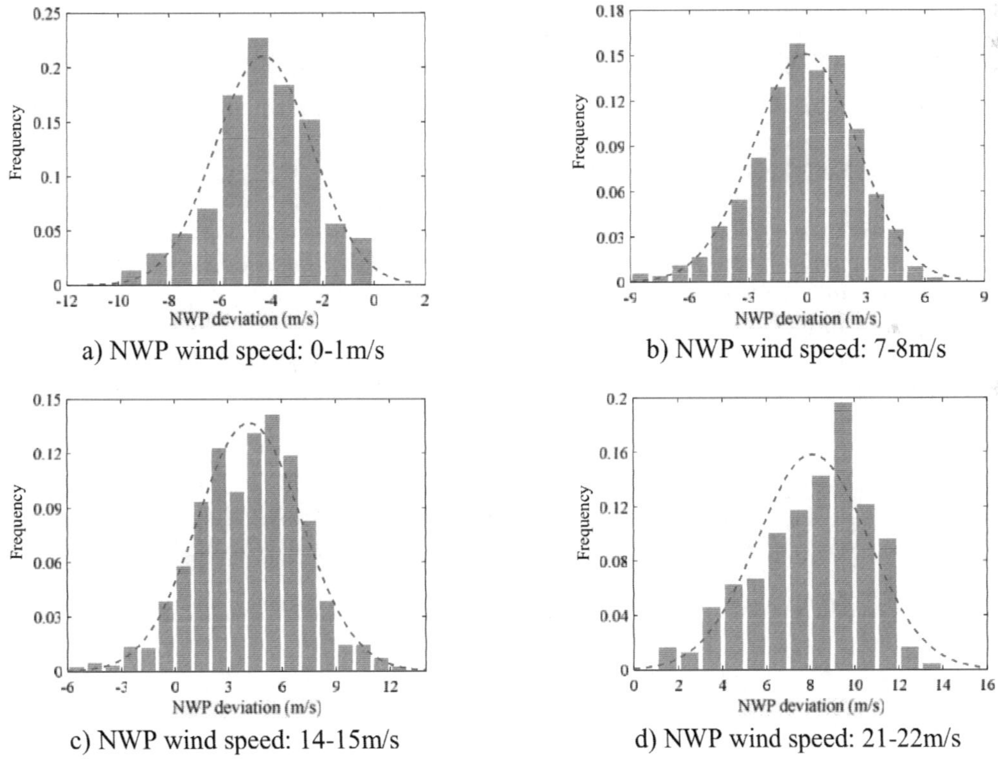

Figure 2. Frequencies of NWP deviation in several wind speed bins

Table 1. Statistics of NWP deviation in several wind speed bins

NWP bin	positive deviation (m/s)		negative deviation (m/s)	
(m/s)	maximum	average	maximum	average
0-1	/	/	9.55	4.30
7-8	6.64	1.97	9.91	2.18
14-15	12.7	4.66	5.97	1.55
21-22	13.2	8.11	/	/

The transfer relationship within the wind speed sequence is the basics for time-series method to predict wind speed. Therefore, the range of NWP deviation can be reduced when the measured wind speed at previous time is also considered as an input label. And the corresponding relationship between inputs and outputs of the prediction correction model will be much clearer under this condition. For example, the frequency of NWP deviation is shown in Fig. 3 when NWP wind speed at time of *t* and the measured wind speed at time of *t-1* are both limited in 7-8m/s. Compared with Fig. 2. b), the frequency of NWP deviation is more concentrated, and the range of NWP deviation is reduced obviously, only between -1 m/s to 1 m/s.

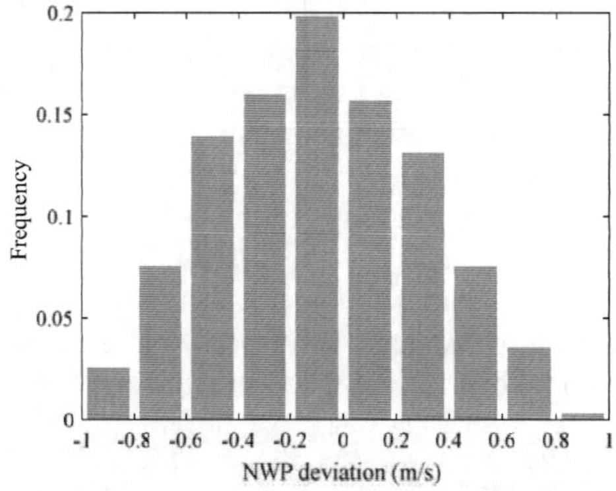

Figure 3. Frequency of NWP deviation
(Wind speed of NWP at time of *t* and HMD at time of *t-1* are both limited in 7~8m/s)

3. The Proposed Wind Speed Prediction Correction Method

This section describes the proposed wind speed prediction correction method construction. The results from section 2 show that even if NWP wind speed is within very small bins, the corresponding relationship between NWP and the measured wind speed at the same time slot is still obscure. Therefore, the time-independent prediction correction methods cannot largely improve the accuracy of NWP. In order to further improve the correction performance, as mentioned before, the impacts passing from the wind speed in the latest time series should be considered. A multi-step-ahead wind speed prediction correction method is proposed in this paper. The main idea of this work is to take the passing effects from wind speed at the previous time slot into account by adding the historical measurement data as another input variable. The proposed model is a rolling prediction correction model, the model and input variables of the model are updated in real-time. In this paper, SVM is employed to validate the proposed modelling strategy, and any AI-based algorithm, such as neural network, relevance vector machine, deep learning, etc. can be easily applied to implement the proposed strategy.

The main modelling process of the proposed wind speed prediction correction method is shown in Fig. 4.

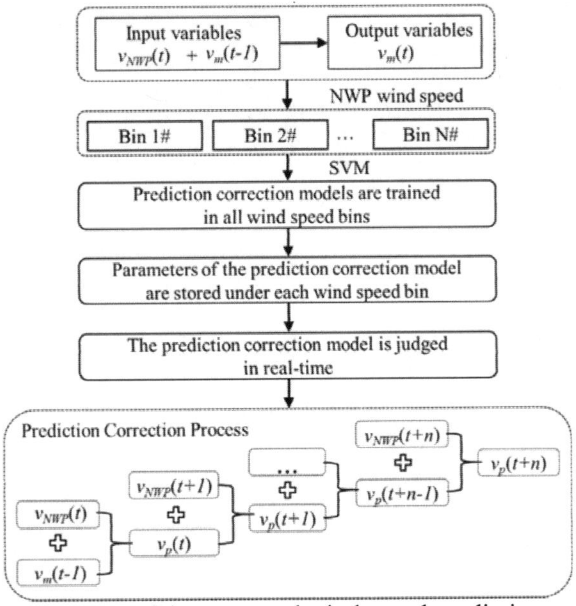

Figure 4. Modelling process of the proposed wind speed prediction correction method

1) To collect the training samples, and to separate the samples into several groups according to NWP wind speed segments. In this case, the segment is 1 m/s for each.

2) To train the model for mapping the relationship between NWP wind speed and the wind speed measurements. NWP wind speed at time of t and the wind speed measurement at time of $t-1$ are all imported as training inputs. And the measured wind speed at time of t is imported as training targets.

3) In this case, SVM is used to establish the wind speed prediction correction model for every wind speed segment, and the model parameters in each wind speed segment are stored to the future correction.

4) The prediction correction model is determined and updated in real-time according to the NWP wind speed. And then the corresponding model is applied to the wind speed prediction correction. Except for the first wind speed correction point, the measured wind speed is substituted by predicted wind speed for input variables during the prediction in the proposed method.

The modelling process of SVM is as follows.

SVM [16] is proposed by Vapnik. This algorithm not only has the ability to deal with nonlinear regression problems, but also has the global optimal solution. It is suitable for solving small sample, nonlinear and high-dimensional pattern recognition problems. To solve the nonlinear regression problem of wind speed prediction, a nonlinear mapping function is used to map the original sample space into a high dimensional space, and then the linear regression method is applied to analysis. The regression linear function is shown as eq. (2).

$$f(x)=\langle w\varphi(x)\rangle+b \tag{2}$$

where, w is the weight vector; $\varphi(x)$ is the nonlinear mapping function; b is the threshold.

So the solution of the regression problem is converted to search for optimal w and b. The expression of mathematical programming problem is shown as eq. (3).

$$\min\left(\frac{1}{2}\|w\|^2+C\sum_{i=1}^{N}\left(\xi_i+\xi_i^*\right)\right)$$
$$s.t.\begin{cases} y_i-w\,\varphi(x_i)-b\le\varepsilon+\xi_i \\ w\,\varphi(x_i)+b-y_i\le\varepsilon+\xi_i^* \\ \xi_i\ge 0 \\ \xi_i^*\ge 0 \end{cases} \tag{3}$$

where, C is the penalty factor to equilibrate the model complexity and error term; ε is the estimated accuracy; ζ_i, ζ_i^* are relaxation variables, whose purpose are to deal with data that cannot be estimated; x_i are the inputs of training samples; y_i are the outputs of training samples.

Gaussian radial basis kernel function is selected as kernel function in this paper, as shown in eq. (4).

$$K(x_i,x)=\exp\frac{\|x_i-x\|}{2\sigma^2} \tag{4}$$

where, σ is the nuclear parameter. The minimization form is transformed into dual form by Lagrange function. Thus, the wind speed prediction model (regression function) can be expressed as eq. (5).

$$f(x)=\sum_{i=1}^{N}\left(\alpha_i-\alpha_i^*\right)K(x_i,x)+b \tag{5}$$

where, α_i, α_i^* are Lagrange multipliers.

4. Case Study

In this section, the NWP and measured wind speed of the wind farm in north China is taken as example to analysis. Temporal resolution of the data is 10 minutes. Data of the first 20 days each month are taken as the train samples, and the rest data of each month are taken as the test samples. Based on two widely acknowledged criteria, three conventional benchmark methods are compared with the performance of the proposed model, and followed by the error analysis for different time horizons.

4.1. Performance Criteria

In order to evaluate the accuracy of wind speed prediction models, two error indexes are used as evaluation criteria. One is root mean square error (RMSE), formulated in eq. (6), to evaluate the overall error. Another is mean absolute percentage error (MAPE), formulated in eq. (7), to evaluate the real-time error.

$$RMSE=\sqrt{\frac{\sum_{t=1}^{N}\left(v_p(t)-v_m(t)\right)^2}{N}} \tag{6}$$

$$MAPE=\frac{100}{N}\sum_{t=1}^{N}\left|\frac{v_p(t)-v_m(t)}{v_m(t)}\right| \tag{7}$$

where, $v_m(t)$ is the measured wind speed at time of t; $v_p(t)$ is the predicted wind speed at time of t; N is the length of prediction.

4.2. Results and Analysis

In this case study, three widely used methods are selected as the benchmark, which are ARMA, SVM and SVM in all wind speed bins (SVM-bins). ARMA is a typical time-series method working well for the ultra-short-time prediction, it is mainly used to validate the performance of 1-4h prediction. SVM is one of mainstream algorithm for establishing the mapping relationship between NWP and the measured wind speed at the same time slot. SVM-bins method is to establish typical SVM model for each wind speed bin, in this case 1 m/s per bin.

Fig. 5 compares the error of raw NWP data, and corrected NWP data by the proposed method and the other three benchmarks. The "NWP" represents original NWP. The "NWP & HMD" represents the results of the proposed wind speed prediction correction model. RMSE and MAPE of the proposed model are compared with those of the original NWP as well as the other three conventional models under different time horizons. The percentage of the accuracy improvement is calculated, and the comparison results to the original NWP and the second-ranking benchmark are shown in Tab. 2 and Tab. 3. It can be seen that, the proposed wind speed prediction correction model has the highest accuracy comparing with the three benchmarks under different time horizons for both evaluation criteria. For 1h and 4 h, ARMA is in the second-ranking of this comparison. Compared with ARMA, the prediction accuracy of the proposed model is increased 8.15% and 1.76% of RMSE and MAPE under 1h time horizon, 8.66% and 0.80% of RMSE and MAPE under 4h time horizon. For 12h, SVM-bins and ARMA are the second-ranking of this comparison when using RMSE and MAPE as the evaluation criterion respectively. The improvement accuracy of the proposed model is 16.67% of RMSE comparing with SVM-bins, 15.26% of MAPE comparing with ARMA under 12h time horizon. For 24h, SVM-bins is in the second-ranking of this comparison. Compared with SVM-bins, the prediction accuracy of the proposed model is increased 13.41% and 20.87% of RMSE and MAPE under 24h time horizon.

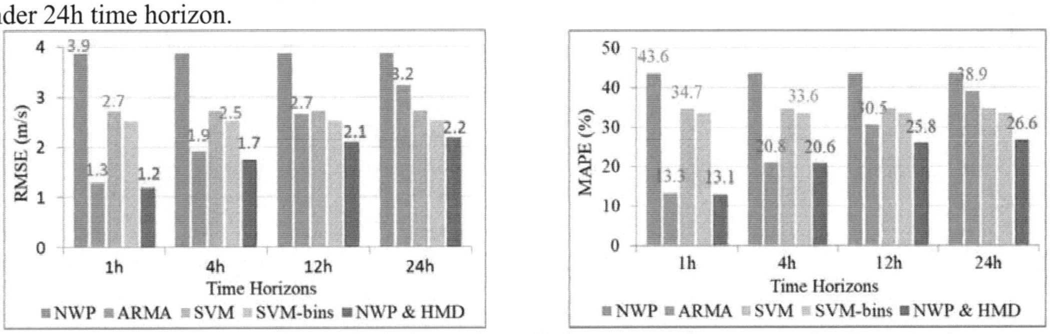

Figure 5. Comparison of wind speed prediction error using two evaluation criteria

Table 2. Improvement compared with original NWP

	1h	4h	12h	24h
RMSE	69.22%	54.91%	45.59%	43.46%
MAPE	70.06%	52.69%	40.81%	39.12%

Table 3. Improvement compared with the second-ranking method

	1h	4h	12h	24h
RMSE	8.15% (ARMA)	8.66% (ARMA)	16.67% (SVM-bins)	13.41% (SVM- bins)
MAPE	1.76% (ARMA)	0.80% (ARMA)	15.26% (ARMA)	20.87% (SVM- bins)

To further validate the adaptability of the proposed method, RMSE in different months are compared for the four models under ultra-short-term and short-term time scales. The results are shown in Fig. 6. It can be seen that in each month the proposed method outperforms the other benchmarks under various time horizons. And, for the ultra-short-term (1 and 4 hours ahead) the performance of the proposed method is similar to that of ARMA at around 1-2m/s and 1.5-2.5m/s; while for the short-term (12 and 24 hours ahead) the proposed method seems to have larger advantages comparing to the rest methods, and the deviation range of the proposed method for 12h and 24h is around 2-3m/s. Also, the degradation of the proposed method from 1h to 24h is not as large as that of ARMA. Fig. 7 shows the accuracy degradation of the proposed method with the increase of the prediction horizons. As shown, the RMSE and MAPE of the proposed method increase sharply before 8-hour time horizon, and stays constant when the time horizon is over 12h.

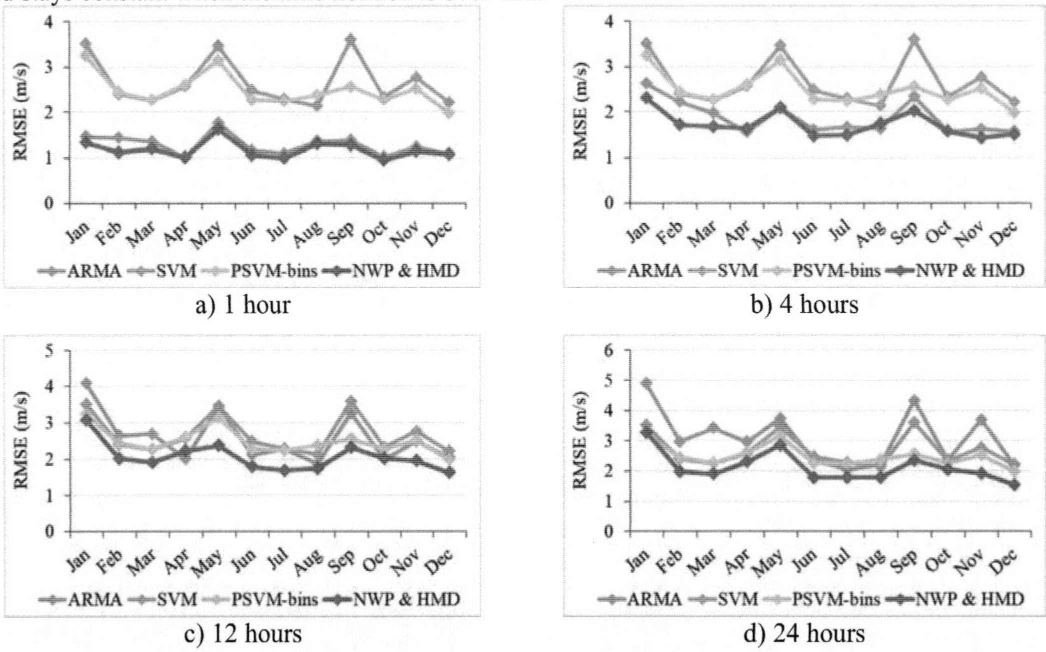

Figure 6. RMSE of four prediction models in different months

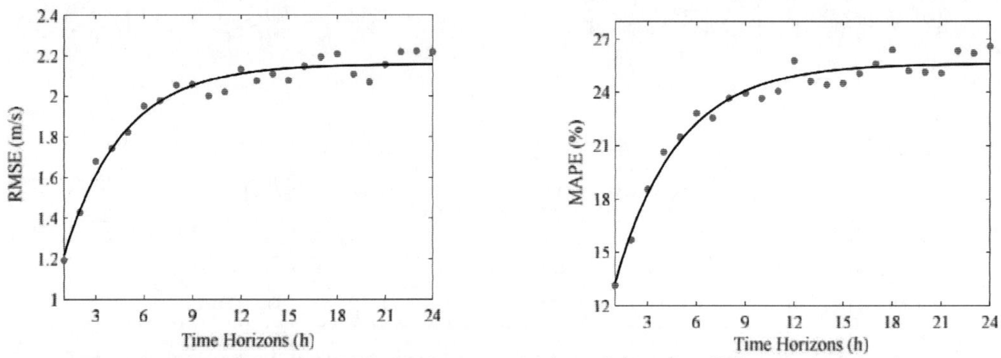

Figure 7. RMSE and MAPE of the proposed model under different time horizons

Deviation of the wind speed prediction can be divided into two parts, one is the statistical deviation, which might be able to eliminate by the statistical method and mapping algorithm to capture its pattern, for example, deviation grows with the wind speed increases; another is the random deviation, which might not be easy to capture its pattern. The correction of wind speed prediction aims to eliminate the statistical deviation and to leave the random deviation to the corrected NWP wind speed. Fig. 8 shows the corrected results for different wind speed bins. For the time horizon of 1h, the RMSE of each wind speed bin roughly ranges from 1m/s - 1.5m/s, which is different from the original NWP deviation pattern. For the rest time horizons, the results show similar but not such apparent features.

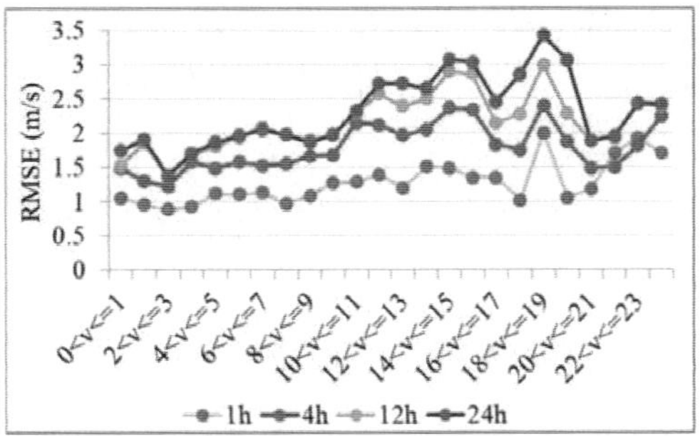

Figure 8. RMSE of the proposed model in different NWP wind speed bins

5. Conclusion

In this paper, a multi-step-ahead wind speed prediction correction method is proposed with consideration of the impacts passing from the wind speed in the latest time series. Both numerical weather prediction and historical measurement data are used as training inputs. One Chinese wind farm in northern part of China is taken as example to validate the proposed method. The conclusions of this paper are as follows.

1) Through the probabilistic analysis, the NWP deviation pattern is not very clear when only the NWP wind speed interval is considered. The actual wind speed varies widely even if the NWP wind speed is within a very small range. The involvement of wind speed at previous time slot enhances the capture of a clearer deviation pattern. This motivates to present a time-dependent method to correct the wind speed prediction in this paper.

2) In this case study, the proposed method can improve the accuracy of NWP wind speed to a large extent at different time horizons. Compared to the second-ranking benchmark method, the proposed method outperforms by 8.15%, 8.66%, 16.67% and 13.41% at the time horizons of 1h, 4h, 12h and 24h in terms of RMSE. Also, the accuracy degradation of the proposed method is not as large as the traditional time-series based method especially.

Acknowledgments

This paper is supported by National Natural Science Foundation of China (Project No. 51376062) and the project of "system development of the optimization and evaluation of the wind farm siting and wind turbine selection" from United Power.

Reference

[1] Du P, Wang J Z, Guo Z H and Yang W D 2017 Research and application of a novel hybrid forecasting system based on multi-objective optimization for wind speed forecasting *Energy Conversion and Management* **150** 90-107

[2] Li H B, Lu Z X, Qiao Y and Zeng P L 2015 Assessment on operational flexibility of power grid with grid-connected large-scale wind farms *Power System Technology* **39** 1672-78

[3] Ren G R, Liu J F, Wan J, Guo Y F and Yu D R 2017 Overview of wind power intermittency: impacts, measurements, and mitigation solutions *Applied Energy* **204** 47-65

[4] Giebel G, Cline J, Frank H, Shaw W, Pinson P, Hodge B.-M, Kariniotakis G, Madsen J and Mo¨hrlen C 2016 Wind power forecasting: IEA Wind Task 36 & future research issues *Journal of Physics: Conference Series* **53** 032042-10

[5] Lazarevska E 2016 Comparison of Different Models for Wind Speed Prediction *IECON 2016 - 42nd Annual Conference of the IEEE Industrial Electronics Society* (IEEE, Piscataway, NJ, USA: Elsevier) 5544-9

[6] Kavasseri R G and Seetharaman K 2009 Day-ahead wind speed forecasting using f-ARIMA models *Renewable Energy* **34** 1388-93

[7] Urata K, Inoue M, Murayama D and Adachi, S 2016 Dynamic Modelling and Very Short-term Prediction of Wind Power Output Using Box-Cox Transformation *Journal of Physics: Conference Series* **744** 012176

[8] Galanis G, Papageorgiou E and Liakatas A 2017 A hybrid Bayesian Kalman filter and applications to numerical wind speed modeling *Journal of Wind Engineering and Industrial Aerodynamics* **167** 1-22

[9] Zhang Y, Sun H X, Guo Y J and Lu Z P 2016 Improved grey model with rolling method for wind power prediction *35th Chinese Control Conference* (IEEE, Piscataway, NJ, USA: Elsevier) 9784-6

[10] Li L, Liu Y Q, Yang Y P, Han S and Wang Y M 2013 A physical approach of the short-term wind power prediction based on CFD pre-calculated flow fields *Journal of Hydrodynamics* **25** 56-61

[11] Wang D Y, Luo H Y, Grunder O and Lin Y B 2017 Multi-step ahead wind speed forecasting using an improved wavelet neural network combining variational mode decomposition and phase space reconstruction *Renewable Energy* **113** 1345-58

[12] Sapronova A, Meissner C and Mana M. 2016 Short time ahead wind power production forecast *Journal of Physics: Conference Series* **749** 012006

[13] Yan J, Gao X L, Liu Y Q, Han S, Li L, Ma X M, Gu C H, Bhakar R and Furong L 2015 Adaptabilities of three mainstream short-term wind power forecasting methods *Journal of Renewable and Sustainable Energy* **7** 053101

[14] Zang H X, Fan L, Guo M, Wei Z N, Sun G Q and Zhang L 2016 Short-Term Wind Power Interval Forecasting Based on an EEMD-RT-RVM Model *Advances in Meteorology* **2016** 8760780

[15] Wang H Z, Li G Q, Wang G B, Peng J C, Jiang H and Liu Y T 2017 Deep learning based ensemble approach for probabilistic wind power forecasting *Applied Energy* **188** 56-70

[16] Vapnik V 1995 *The nature of statistical learning theory* New York: Springer-Verlag

IOP Conf. Series: Journal of Physics: Conf. Series **926** (2017) 012008 doi:10.1088/1742-6596/926/1/012008

Condition monitoring of a wind turbine doubly-fed induction generator through current signature analysis

Estefania Artigao[1], **Andres Honrubia-Escribano**[1,2], **Emilio Gomez-Lazaro**[1,2]

[1] Renewable Energy Research Institute, Paseo de la Investigacion 1, 02071, Albacete, SPAIN
[2] EDII University of Castilla – La Mancha, Avenida de España s/n, 02071, Albacete, SPAIN

E-mail: estefania.artigao@uclm.es

Abstract. Operation and maintenance (O&M) of wind turbines is recently becoming the spotlight in the wind energy sector. While wind turbine power capacities continue to increase and new offshore developments are being installed, O&M costs keep raising. With the objective of reducing such costs, the new trends are moving from corrective and preventive maintenance toward predictive actions. In this scenario, condition monitoring (CM) has been identified as the key to achieve this goal. The induction generator of a wind turbine is a major contributor to failure rates and downtime where doubly-fed induction generators (DFIG) are the dominant technology employed in variable speed wind turbines. The current work presents the analysis of an in-service DFIG. A one-year measurement campaign has been used to perform the study. Several signal processing techniques have been applied and the optimal method for CM has been identified. A diagnosis has been reached, the DFIG under study shows potential gearbox damage.

1. Introduction

Wind energy has been renowned as the fastest growing renewable energy source during the past years, with a global installed wind capacity of nearly 490 GW by the end of 2016, of which, around 15 GW are located offshore [1]. European countries are the leading players in the offshore wind energy market [2].

More modern and larger wind turbine generators are under continuous development. These exhibit more faults when compared to smaller ones [3], which becomes critical offshore [4]. Under this framework, operation and maintenance (O&M) [4] is key to improve reliability and availability. O&M activities represent between 25-35% of the total expenditure of a wind farm project [5], hence, it is mandatory to optimise them in order to reduce costs. There are three common types of maintenance strategies: preventive, corrective and predictive [6]. Downtime is avoided in the first strategy, while unnecessary resources and expenses may take place. For the second one, only the required resources are used but it can result in excessive downtime. New trends are moving from these two towards predictive maintenance, where condition monitoring (CM) determines the ideal balance between preventive and corrective actions [7], improving reliability and availability while reducing costs [8].

The generator is one of the major contributors to failure rates and downtime of wind turbines, together with the gearbox and the drive train system [9]. Doubly-fed induction generator (DFIG) is the dominant technology employed for variable speed wind turbines [10]. The most common faults of DFIGs are related to bearings (between 20% and 38%), stator (around 30%) and rotor (nearly 10%) [11,12]. Induction machine's condition monitoring techniques include temperature, chemical and wear, mechanical vibration, electrical current, and power, electrical discharge and artificial intelligence, where electrical signature analysis is the most common one [13].

2. Methods

Electrically-based condition monitoring techniques include: current, voltage, instantaneous power and flux analysis, where stator current analysis is the most common of these techniques. It is based on the principle that each fault has its own effect on the current spectra [10]. Several signal processing methods can be applied to electrically-based techniques towards fault identification, typically classified into time-domain, frequency-domain and time-frequency [13].

In the present work, frequency-domain analysis will be applied on a real operating DFIG wind turbine via Fast Fourier Transform (FFT) to both stator and rotor instantaneous currents. This technique transforms the waveform signal from the time domain to the frequency domain. It is widely used for fault diagnosis because the variations of certain harmonic components in the frequency spectrum of a signal can be related to a specific fault type [14]. The formulae that provides the frequency components associated to various types of faults have been identified and demonstrated for different kinds of induction machines. A summary of such formulae was presented in [15, 16] by the authors. Particularly, the equation that provides frequency components related to gearbox damage is defined as (1):

$$f_{gb} = f_s \pm m\left[\left(\frac{f_s}{G_r p/2}\right)\right] \tag{1}$$

Where f_s is the supply frequency, m is the harmonic order, G_r is the gearbox ratio and p is the number of pole pairs.

Current Park Vector is another approach for fault diagnosis of electric machines. It is a fairly recent technique that can successfully diagnose rotor faults, inter-turn stator faults and unbalanced supply voltage and mechanical load-misalignment [17]. The instantaneous stator line currents are transformed into the Parks vector as per equations (2) and (3). An undamaged machine theoretically shows a perfect circle, and an unbalance one results in an elliptic representation.

$$I_d = \left(\sqrt{\frac{2}{3}}\right) I_a - \left(\sqrt{\frac{1}{6}}\right) I_b - \left(\sqrt{\frac{1}{6}}\right) I_c \tag{2}$$

$$I_q = \left(\sqrt{\frac{1}{2}}\right) I_b - \left(\sqrt{\frac{1}{2}}\right) I_c \tag{3}$$

Actual data from operating wind turbines is seldom presented. Much work has been published on modelled data, simulations and laboratory benches [18–22] but there is a lack of field measurement campaigns for current signature analysis. In the present work, an in-service DFIG wind turbine with potential gearbox faults has been analysed through electrically-based techniques. The data for the case study has been provided by the Spanish company Ingeteam Service S.A., specialising in power converters, generators, turbine controllers, CM systems and SCADA systems.

IOP Conf. Series: Journal of Physics: Conf. Series **926** (2017) 012008 doi:10.1088/1742-6596/926/1/012008

3. Results

The reported wind turbine has a doubly-fed induction generator with the following characteristics, Table 1:

Table 1. Main characteristics of the DFIG under study.

Rated Power	1545 kW
Pole pairs	3
Synchronous Speed	1000 rpm
Supply frequency	50 Hz
Nominal stator voltage	12000 V
Nominal rotor voltage	690 V

A test case with the wind turbine at 90% of its rated nominal power operating at super-synchronous speed with 8% slip has been chosen to illustrate the study. The waveforms of the stator and rotor currents are shown in Figure 1 and their main parameters extracted in Table 2.

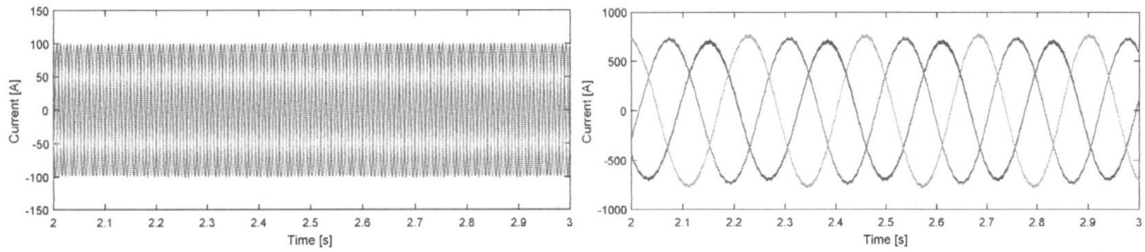

Figure 1. Stator (left) and rotor (right) currents of the DFIG under super-synchronous speed operation (1 s zoom).

Table 2. Current parameters of the DFIG under study.

	Stator rms current [A]	Stator freq [Hz]	Rotor rms current [A]	Rotor freq [Hz]	slip
Phase a	69.5792		490.7019		
Phase b	70.1309	50.0916	511.4118	4.45	-0.08
Phase c	69.2850		535.1995		

The generator slip is calculated dividing the supply frequency (main frequency of the stator) by the main frequency of the rotor, obtaining 0.08. The negative sign is the convention for super-synchronous speed.

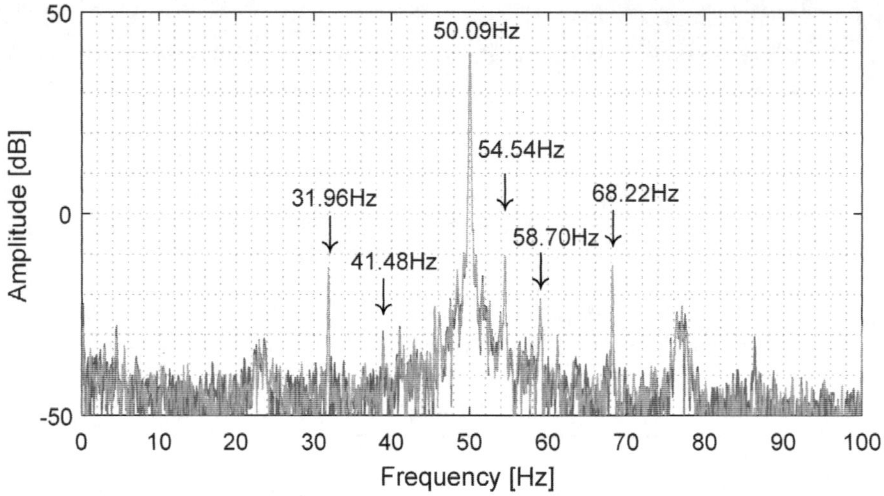

Figure 2. FFT of stator currents.

3.1. Spectral Analysis of the Stator Currents

FFT is now applied to the stator currents, shown in Figure 2, in order to carry out fault frequency search.

The supply frequency is clearly seen in the spectrum with the highest amplitude. Two pair of peaks that are usually present in the spectrum due to the geometry of the induction generator itself are those related to broken rotor bars (41.48 Hz – 58.70 Hz) and rotor unbalance (31.96 Hz – 68.22 Hz). The sum of the stator main frequency (or supply frequency) and the rotor main frequency (4.45 Hz) can also be identified in the spectrum (54.54 Hz).

The mere presence of these components might not imply a fault. It has been reported that less than 25 dB difference must exist between the supply frequency and broken rotor bar components in order to be able to diagnose such fault [11, 23]. Similarly occurs for rotor unbalance components. A minimal degree of air gap is always introduced during manufacturing or assembly of the induction machine, which modulates the current at the rotational speed [19]. When the air gap degree increases, the rest of the sub-harmonics related to rotor unbalance appear, indicating the presence of the fault. In this case, only the first pair can be seen.

Differently occurs with gearbox related components, which shall not appear unless a fault is present. Table 3 summarizes the calculations for the frequency components as per Eq. 1 applied to the wind turbine under study.

Table 3. Gearbox damage related frequency components for the DFIG under study.

m	1	2	3	4	5	6	7
Freq [Hz]	49.58	49.08	48.57	48.06	47.56	47.05	46.56
	50.60	51.10	51.61	52.11	52.62	53.13	56.65

Zoom is applied around the supply frequency (Figure 3 left), where further peaks arise, in order to identify the potential fault frequency components calculated in Table 3. Three pairs corresponding to $m = \pm 2, 3, 5$ can be seen in the spectrum. For a better understanding of the results, the frequency components are represented as a function of m, (Figure 3 right). The presence of f_{gb} components indicate gearbox-related fault. This diagnose has been confirmed by the data owners, Ingeteam.

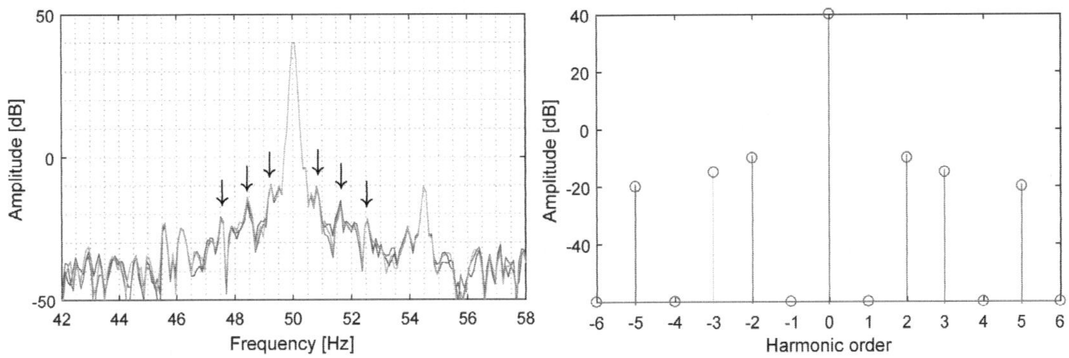

Figure 3. Zoom around supply frequency of the stator currents FFT (left) and gearbox related components by harmonic order (right).

3.2. Spectral Analysis of the Rotor Currents

The rotor shows electrical unbalance, which can be appreciated from the rotor currents previously presented (Figure 1 right) and the rotor rms values extracted in Table 2.

The spectrum analysis of the rotor currents is performed (Figure 4). Frequency components related to rotor asymmetry can be seem in the rotor spectra, corresponding to $3sf$ and $5sf$ respectively, pointing out the rotor electrical unbalance.

3.3. Current Park Vector Analysis

Finally, Current Park vector is applied, following Eqs. 2 and 3 previously defined. The results are shown in Figure 5.

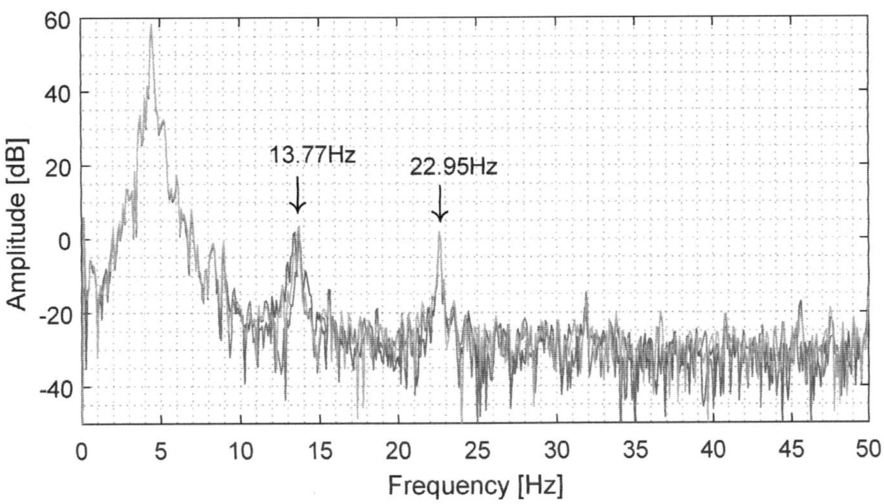

Figure 4. FFT of rotor currents.

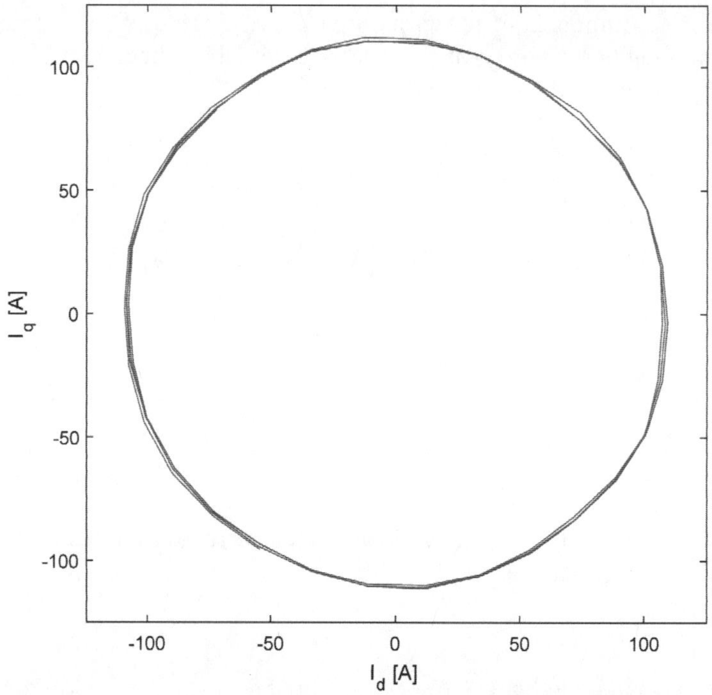

Figure 5. Current Park Vector of the DFIG under study.

The plot shows a perfect circle with certain shift, confirming the diagnose for electrical rotor unbalance and the absence of mechanical eccentricity. Park Vector approach is however unable by definition to validate the diagnose for a potential gearbox damage, as obtained by the spectral analysis of the stator currents.

4. Conclusions

A brief introduction on the state of the art of condition monitoring of doubly-fed induction generators of wind turbines has been given. The motivation being to optimise O&M activities by reducing costs and increasing reliability and availability.

The basis of current signature analysis and its application on DFIGs have been presented, including stator and rotor currents analysis as well as Current Park Vector approach, which have been applied on field measurements from an operating wind turbine. Faulty components related with gearbox damage have been identified on the spectral analysis of stator currents, and electrical rotor unbalance from the rotor analysis.

Both current and rotor spectral analysis show potential with regards to Condition Monitoring development for DFIG-based wind turbines, whereas Current Park Vector appears to provide scarce information for this purpose.

5. Acknowledgements

This project has received funding from the European Union Horizon 2020 research and innovation programme under the Marie Sklodowska-Curie grant agreement No. 642108, AWESOME Project. Special acknowledgement to Ingeteam Service S.A. (part of the AWESOME Project Consortium) who provided the wind turbine data.

6. References

[1] E. Warner, "Global wind report. Annual market update," Global Wind Energy Council, Tech. Rep., 2015.

[2] E. Zountouridou, G. Kiokes, S. Chakalis, P. Georgilakis, and N. Hatziargyriou, "Offshore floating wind parks in the deep waters of Mediterranean Sea," *Renewable and Sustainable Energy Reviews*, vol. 51, pp. 433–448, 2015.

[3] F. Spinato, P. J. Tavner, G. J. W. Bussel, and E. Koutoulakos, "Reliability of wind turbine subassemblies," *IET Renewable Power Generation*, vol. 3, no. 4, pp. 387–401, 2008.

[4] B. Hahn, M. Durstewitz, and K. Rohrig, "Reliability of wind turbines," in *Wind Energy*, 2007, pp. 329–332.

[5] M. I. Blanco, "The economics of wind energy," *Renewable and Sustainable Energy Reviews*, vol. 13, no. 6, pp. 1372–1382, 2009.

[6] F. P. G. Márquez, A. M. Tobias, J. M. P. Pérez, and M. Papaelias, "Condition monitoring of wind turbines: Techniques and methods," *Renewable Energy*, vol. 46, pp. 169–178, 2012.

[7] W. Qiao, P. Zhang, and M.-Y. Chow, "Condition monitoring, diagnosis, prognosis, and health management for wind energy conversion systems," *IEEE Transactions on Industrial Electronics*, vol. 62, no. 10, pp. 6533–6535, 2015.

[8] P. Tchakoua, R. Wamkeue, M. Ouhrouche, F. Slaoui-Hasnaoui, T. A. Tameghe, and G. Ekemb, "Wind turbine condition monitoring: State-of-the-art review, new trends, and future challenges," *Energies*, vol. 7, no. 4, pp. 2595–2630, 2014.

[9] J. Carroll, A. McDonald, and D. McMillan, "Failure rate, repair time and unscheduled O&M cost analysis of offshore wind turbines," *Wind Energy*, vol. 19, pp. 1107–1119, 2015.

[10] C. A. Rieg, C. J. Smith, and C. J. Crabtree, "Monitoring wind turbine loading using power converter signals," in *Journal of Physics Conference Series 749*, 2016.

[11] S. Nandi, H. A. Toliyat, and X. Li, "Condition monitoring and fault diagnosis of electrical motors - A review," *IEEE Transactions on Energy Conversion*, vol. 20, no. 4, pp. 719–729, 2005.

[12] M. E. H. Benbouzid, "A review of induction motors signature analysis as a medium for faults detection," *IEEE Transactions on Industrial Electronics*, vol. 47, no. 5, pp. 984–993, 2000.

[13] C. Sheng, Z. Li, L. Qin, Z. Guo, and Y. Zhang, "Recent progress on mechanical condition monitoring and fault diagnosis," *Procedia Engineering*, vol. 15, pp. 142–146, 2011.

[14] Y. Amirat, M. E. H. Benbouzid, E. Al-Ahmar, B. Bensaker, and S. Turri, "A brief status on condition monitoring and fault diagnosis in wind energy conversion systems," *Renewable and Sustainable Energy Reviews*, vol. 13, no. 9, pp. 2629–2636, 2009.

[15] E. Artigao, A. Honrubia-Escribano, and E. Gomez-Lazaro, "Current signature analysis to monitor DFIG wind turbine generators: A case study," *Renewable Energy*, 2017.

[16] E. Artigao, A. Honrubia, and E. Gomez, "Advanced diagnosis of doubly fed induction generators for wind turbines," in *Wind Energy Science Conference*, 2017.

[17] H. Douglas, P. Pillay, and P. Barendse, "The detection of interturn stator faults in doubly-fed induction generators," in *Fourtieth IAS Annual Meeting. Conference Record of the 2005 Industry Applications Conference*, 2005, pp. 1097–1102.

[18] C. J. Crabtree, S. Djurovic, P. J. Tavner, and A. Smith, "Condition monitoring of a wind turbine DFIG by current or power analysis," in *Power Electronics, Machines and Drives (PEMD), 5th IET International Conference.* IET, 2010, pp. 1–6.

[19] S. Williamson and S. Djurovic, "Origins of stator current spectra in DFIGs with winding faults and excitation asymmetries," *IEEE International Electric Machines and Drives Conference, IEMDC'09*, vol. 9, pp. 563–570, 2009.

[20] S. Djurovic, C. J. Crabtree, P. J. Tavner, and A. Smith, "Condition monitoring of wind turbine induction generators with rotor electrical asymmetry," *IET Renewable Power Generation*, vol. 6, no. 4, pp. 207–216, 2012.

[21] A. Stefani, A. Yazidi, C. Rossi, F. Filippetti, D. Casadei, and G.-A. Capolino, "Doubly fed induction machines diagnosis based on signature analysis of rotor modulating signals," *IEEE Transactions on Industry Applications*, vol. 44, no. 6, pp. 1711–1721, 2008.

[22] A. Yazidi, H. Henao, G. Capolino, D. Casadei, F. Filippetti, and C. Rossi, "Simulation of a doubly-fed induction machine for wind turbine generator fault analysis," in *5th IEEE International Symposium on Diagnostics for Electric Machines, Power Electronics and Drives (SDEMPED)*, 2005, pp. 1–6.

[23] G. Acosta, C. Verucchi, and E. Gelso, "A current monitoring system for diagnosing electrical failures in induction motors," *Mechanical Systems and Signal Processing*, vol. 20, no. 4, pp. 953–965, 2006.

On the use of high-frequency SCADA data for improved wind turbine performance monitoring

E Gonzalez[1], B Stephen[2], D Infield[2], J J Melero[1]
[1]CIRCE - Universidad de Zaragoza, C/ Mariano Esquillor Gómez, 15, 50018 Zaragoza, Spain
[2]Department of Electronic and Electrical Engineering, University of Strathclyde, Glasgow, UK
E-mail: egonzalez@fcirce.es

Abstract. SCADA-based condition monitoring of wind turbines facilitates the move from costly corrective repairs towards more proactive maintenance strategies. In this work, we advocate the use of high-frequency SCADA data and quantile regression to build a cost effective performance monitoring tool. The benefits of the approach are demonstrated through the comparison between state-of-the-art deterministic power curve modelling techniques and the suggested probabilistic model. Detection capabilities are compared for low and high-frequency SCADA data, providing evidence for monitoring at higher resolutions. Operational data from healthy and faulty turbines are used to provide a practical example of usage with the proposed tool, effectively achieving the detection of an incipient gearbox malfunction at a time horizon of more than one month prior to the actual occurrence of the failure.
Keywords: Operation & Maintenance, Wind Turbine, Power Curve, Performance Monitoring, SCADA, High-frequency data, Fault Detection

1. Introduction

High wind farm (WF) operation and maintenance (O&M) costs are a major concern today. The maintenance requirements due to wind turbine (WT) failures, together with poor site accessibility, can lead to rates up to 30% of the levelised cost of energy (LCOE) of an offshore WF [1]. Unscheduled maintenance activities not only increase the O&M costs but also have a strong impact on downtime and thus on annual energy production. There is therefore a need to improve the WT O&M phase.

Early detection of failures is desired to reduce O&M costs while improving reliability. To this end, the use of condition monitoring systems (CMS) is required. Although numerous commercial options and techniques are already available [2, 3], there are still limitations for a wide deployment. Some CMS might outstrip the expense of the required additional equipment, but may also exhibit high rates of false positive alarms while the diagnosis is dedicated to a unique component or assembly rather than system-wide [2, 4]. WT condition monitoring based on the use of data from the supervisory control and data acquisition (SCADA) system [5] is increasingly seen as a cost-effective and promising approach, as SCADA data is available at no additional cost. Within this approach, monitoring WT performance is fundamental, since it is undoubtedly the main feature characterising WT overall operation. Component malfunction may degrade the energy conversion efficiency leading to performance deviations. Performance monitoring can thus play a useful role in global condition monitoring.

Generally, performance monitoring is synonymous with power curve monitoring [6]. The most widely used approach consists on modelling a reference power curve, based on operational data under normal operating conditions (WT normal performance), together with confidence intervals [7] for discriminating between normal and abnormal operation over time. Deviations from this reference, beyond these thresholds, denote a fault in the energy extraction process, e.g. due to incipient failures of components [8]. Although important research has been conducted regarding WT power curve modelling, little emphasis has been placed on the estimation of the related confidence limits.

Both parametric [9, 10, 11, 12] and non-parametric methods [9, 13, 14, 15, 16] have been studied to produce a reference for WT normal performance, based on operational data from the SCADA system. While parametric models fit a functional form to the data by changing one or several parameters, non-parametric methods infer a functional form and do not need to make any assumption about the data. Recent research has highlighted the accuracy of non-parametric and multivariate methods [16] for WT power curve modelling. While most contributions rely on univariate modelling (relating wind speed and WT power output), multivariate models allow to take into account the natural variability of WT performance, greatly influenced by several phenomena.

With regard to the data used, the majority of studies and commercial applications are based on the use of 10-minute aggregated data. As stated by Yang *et al.* [3], this low frequency resolution negatively affects the diagnosis and prognosis capabilities. Indeed, this aggregation may hide short-lived events. High-resolution SCADA data instead, should allow dynamic turbine behaviour to be identified with higher fidelity and thus improve detection capabilities [13, 17, 18].

Finally, as previously mentioned, the assessment of WT performance thresholds has not been extensively studied. The estimation of these confidence intervals is directly linked to power curve modelling related uncertainty. Indeed, WT performance can significantly vary under normal operating conditions [19, 20, 21]. Power curve modelling uncertainty should account for this normal variability to ensure effective abnormal behaviour detection. Even though non-parametric approaches can accurately represent WT performance, the assessment of their related uncertainty is not always straightforward. State-of-the-art methods rely on the assumption of a normal distribution of the modelling error [9, 22]. Recent work published by the authors has shown the heteroscedastic nature of the model residuals [18], that is, characterised by a non-constant variance across the different wind speed regimes. As a result, the assumption of a normal distribution of the modelling error may lead to a loss of detection capabilities and its validity should be questioned. WT normal performance thresholds should rather vary across the different wind speed regimes. This paper introduces the use of Quantile Regression Forests [23, 24] and high-frequency SCADA data for modelling WT normal performance together with its related uncertainty, for the purpose of performance monitoring.

The paper is organised as follows. Section 2 details the suggested approach for WT performance monitoring, with four alternatives to assess WT performance and its related uncertainty. In Section 3, the methodology is first tested in a case study, consisting of a healthy WT; modelling accuracy and detection effectiveness are compared for the different modelling techniques and for SCADA data of high and low frequency data. The best approach is then applied to concurrently monitor healthy and faulty turbines for detection purposes. Advantages and shortcomings are discussed in Section 4, together with future work.

2. Methodology

The suggested methodology for WT performance monitoring includes several phases as one can see in Figure 1. The four phases can be summarised as data filtering, performance modelling and assessment of its related uncertainty, then abnormal performance detection and root-cause diagnosis. Each phase is detailed subsequently.

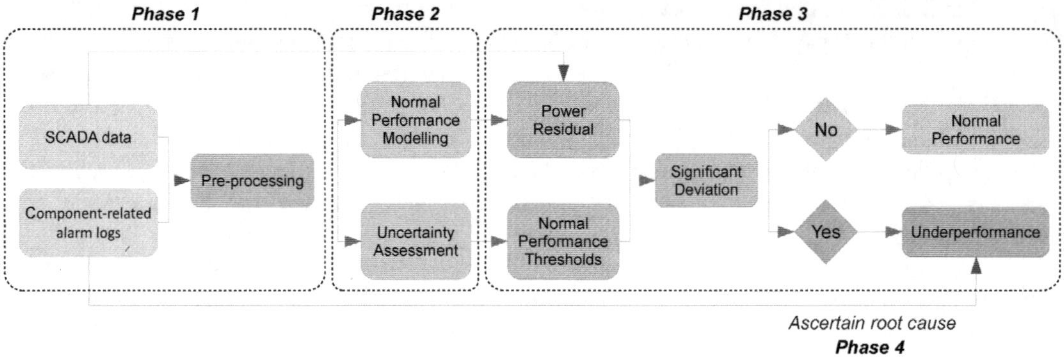

Figure 1. General framework for wind turbine performance monitoring.

During ***Phase 1***, the historical SCADA data used for normal performance modelling should be filtered to ensure faultless representation of WT operation. The pre-processing is addressed in two steps. First, the alarm logs also recorded by the SCADA system are integrated to segregate operational data and flag any abnormal event acknowledged by the system. The SCADA logs are automatically translated into component-related information, based on a modernised WT taxonomy [25]. The interested reader may look in [26] for more information. This way, the SCADA data can be flagged every time the WT was experiencing an issue related to a specific component, to the grid or to extreme weather events; the flagged periods can be then discarded. Since some problems might not be acknowledged by the SCADA system, further actions may be required. In this second step, further abnormal performance filtering is applied based on a multivariate curve approach and machine characteristics, similarly as in [20]. Pitch angle and rotor speed data allow clear identification of the different operating regimes and therefore normal operation, as one can see in Figure 2. All figures presented here are normalised for confidentiality reasons.

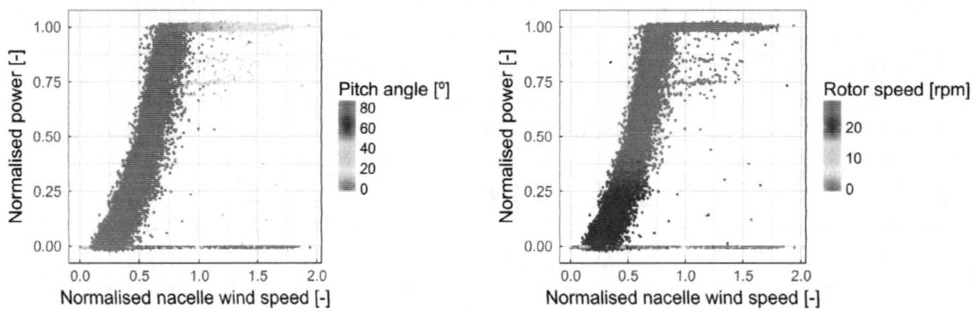

Figure 2. Scatter WT power curve varying depending on blade pitch angle (left) and rotor speed (right).

Phase 2 includes both WT performance modelling and assessment of its related uncertainty. In this paper, four different models are considered for comparison purposes; they are summarised subsequently.

- *Method of bins*

  The method of bins (MOB) is the current industry practice for WT performance testing, as outlined in [27, 28]. It relies on the reduction of the dataset into mean values per wind speed intervals, called bins. For each bin, centred on a multiple of 0,5 m/s, the wind speed and power output are reduced to the calculated mean. Based on these bins and corresponding mean values it is then possible to predict the power output given the wind speed. In this paper, a power curve is obtained based on the nacelle wind speed and the power output. Also, as detailed in [28], data normalisation is applied to the nacelle wind speed to account for the air density effect.

  Similarly to the mean, the standard deviation of the power is calculated per bin to assess the uncertainty related to predictions, as illustrated in Figure 3. This model uncertainty is directly related to the statistical variation of the power output, as defined in [28].

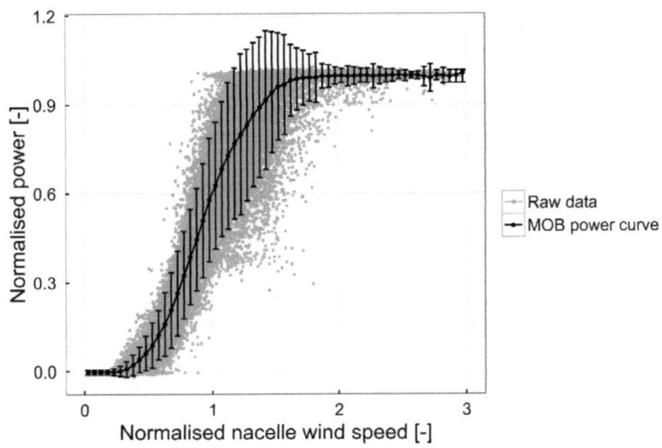

Figure 3. Power curve and related uncertainty derived from the MOB.

- *k-Nearest Neighbours*

  The k-Nearest Neighbours (kNN) is a non-parametric machine learning tool that predicts a new sample using the k closest samples from the training dataset [29]. A multivariate approach is considered here; the nacelle wind speed, ambient temperature, pitch angle and rotor speed data are used as inputs to predict WT power output. Although considering wind direction data as an additional model input may improve WT power prediction [16], this was discarded in this paper due to a lack of data availability. Nevertheless, WT directional behaviour, including wake affected performance, is considered to be inherent in the WT operational data.

  The value of k, characterising the k-Nearest Neighbours, is selected according to the minimum root mean square error (RMSE) from the testing dataset.

  Uncertainty related to the kNN predictions is assessed based on the state-of-the-art practice [9], that assumes a normal distribution of the training error with mean μ_{train} and standard deviation σ_{train}. As a consequence, the same distribution is considered across the different wind speed regimes. Thresholds for normal performance can be derived from control limits (CL) as defined in Equation 1. The parameter η can be tuned iteratively to attain a level of 95% confidence based on the training dataset.

$$CL = \mu_{train} \pm \eta \frac{\sigma_{train}}{\sqrt{N_{train}}} \tag{1}$$

- *Random Forests*

  Random Forests (RF) are also a non-parametric machine learning method built upon multiple regression trees [30]. A conceptual diagram of the algorithm is illustrated in Figure 4, where the left-hand side corresponds to the training phase and the right-hand side to the testing/predicting phase. A single regression tree (e.g. $tree_1$, $tree_2$, or $tree_T$) usually splits the training output values according to conditions of the input values. Different conditions of these input values are represented as branches, while the target values are represented as leaves, corresponding to light blue shaded nodes in Figure 4. RF are considered an ensemble method since they operate by randomly constructing multiple regression trees, avoiding the overfitting problems often related to the use of a single regression tree. As concerns new predictions, RF predict new samples by averaging the individual predicted values from the regression trees, according to conditional inputs; the single predicted values from each tree are highlighted in dark blue in Figure 4.

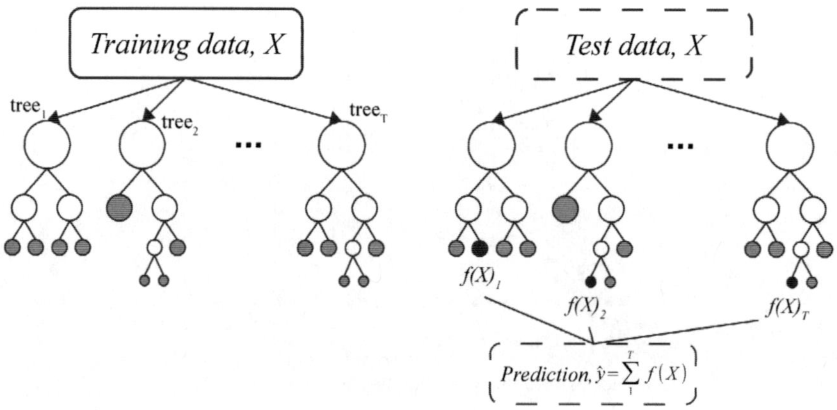

Figure 4. Conceptual diagram of the RF algorithm. Adapted from [31].

Similarly to the previous method, in this paper a power curve is obtained using the nacelle wind speed, ambient temperature, pitch angle and rotor speed data as inputs.
Likewise, uncertainty related to the RF predictions is assessed by the use of control limits derived from the assumed normal distribution of the training error (Equation 1).

- *Quantile Regression Forests*

  As can be seen on the right side of Figure 4, WT power output is predicted from RF as a conditional mean of multiple trees responses. One can therefore easily think that considering the full conditional distribution of the response variable rather than only the mean, would allow to properly address WT power variability under similar input conditions [24]. Quantile Regression Forests (QRF) are based on this principle, as a natural generalisation of the RF algorithm. As a result, quantile regression provides information about the spread of the response variable, according to different levels of desired confidence. The slight variation between the classic RF algorithm and the QRF relies on the main predicted value; the first predicts the response based on the conditional mean whereas the latter predicts the response based on the median of the conditional distribution, corresponding to a level of confidence of 50%. To the best of the author's knowledge, no quantile regression has been applied before for WT performance assessment. In this paper, the WT power curve is modelled by the use of QRF with nacelle wind speed, ambient temperature, pitch angle and rotor speed as input data. Four prediction intervals according to different levels of confidence (80%, 85%, 90% and 95%) are used .

IOP Conf. Series: Journal of Physics: Conf. Series **926** (2017) 012009 doi:10.1088/1742-6596/926/1/012009

As illustrated in Figure 1, ***Phase 3*** consists in the detection of abnormal performance. The difference between the modelled and actual power production is compared to the normal performance thresholds to distinguish between abnormal and expected behaviour. The present work focuses on the detection of WT underperformance. Over and underperformance are here considered to be independent, rather than a common detection as abnormal performance.

Finally, ***Phase 4*** is defined as the diagnosis phase. Once an event of underperformance is detected, the alarms registered by the SCADA system at the same time might be helpful for ascertaining the root cause of the detected deviation. Underperformance events may thereby be attributed to a component-related root cause.

3. Results and discussion

3.1. Data description

SCADA data from an operating WF was used to evaluate the power curve modelling accuracy and the detection capabilities of the described methodology. Two different data time resolutions were considered. First, real measurements recorded at a 4-second resolution, so-called high-frequency data. Second, 10-minute aggregated signals, the current industry practice. One full year of data was taken from three selected turbines. The turbines, of the same type, are located next to each other, at similar altitudes and terrain complexity, hence exposed to similar local conditions. Two of the turbines did not register any failure during the available year, whereas the third registered a major gearbox failure, as indicated in the documented information provided by the WF operator. No information about the specific failure mode was reported. A major failure is understood by the considered operator as entailing either a component repair or replacement.

3.2. Power curve modelling accuracy

High-frequency data from the two healthy turbines (WT1 and WT2) was here used to compare the accuracy of the models described in Section 2. The first month was used to train and test the models, all using a 10-fold cross-validation, by randomly dividing the observations into ten-folds of approximately equal sizes. Modelling accuracy was evaluated by the mean absolute error (MAE) and the root mean squared error (RMSE). Results for the high-frequency data are summarised in Table 1.

Table 1. Performance metrics for power curve modelling accuracy.

	WT1		WT2	
	MAE (%)	RMSE (%)	MAE (%)	RMSE (%)
MOB	4.40	7.22	4.11	6.76
kNN	2.65	5.00	2.66	4.83
RF	1.39	2.65	1.36	2.52
QRF	2.14	4.44	2.03	4.07

In agreement with the literature, all non-parametric models are highly accurate in representing WT normal performance, and the RF outperforms in comparison. The MOB presents higher errors, especially in terms of the RMSE. The univariate and simplistic approach is not able to take the natural variability of WT performance into consideration. The difference between the QRF and the RF, relying on the consideration of the median over the mean, is translated into a slightly significant variation in the prediction accuracy.

3.3. Uncertainty assessment

As detailed in Section 2, the modelling related uncertainty was used to build normal performance thresholds for each method. The model error distribution together with its related normal performance thresholds is depicted in Figures 5 to 8.

Although, the MOB appears to be less accurate in normal performance modelling, the data reduction to a mean and standard deviation per bin accounts for the heteroscedastic nature of the model residual, that is, the variance is not the same for each bin; it therefore produces variable control limits across the different wind speed regimes. The wrong assumption for the uncertainty related to both the kNN and the RF is confirmed in Figure 6 and Figure 7. While the combined distribution may resemble a normal distribution, notwithstanding the high kurtosis, the resulting control limits do not take into account the heteroscedasticity of the residual. This undoubtedly results in a loss of the detection capabilities. Finally, the QRF offers a very flexible way to assess WT performance uncertainty, while having a reasonable modelling accuracy.

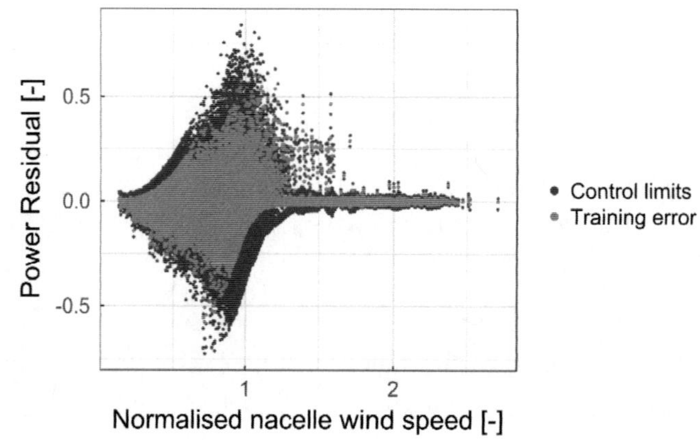

Figure 5. Training error (grey) and related control limits (blue) from the MOB varying depending on the wind speed.

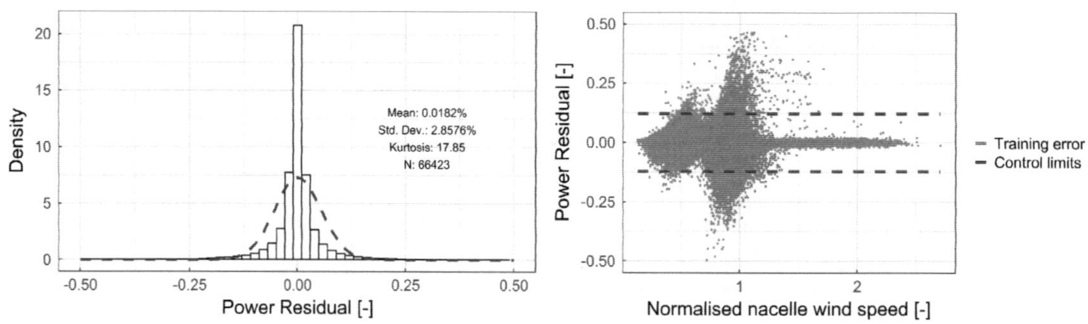

Figure 6. Training error statistics and distribution from kNN model (left side); the related normal distribution is illustrated in blue. Training error (grey) and related control limits (blue) from kNN model varying depending on the wind speed (right side).

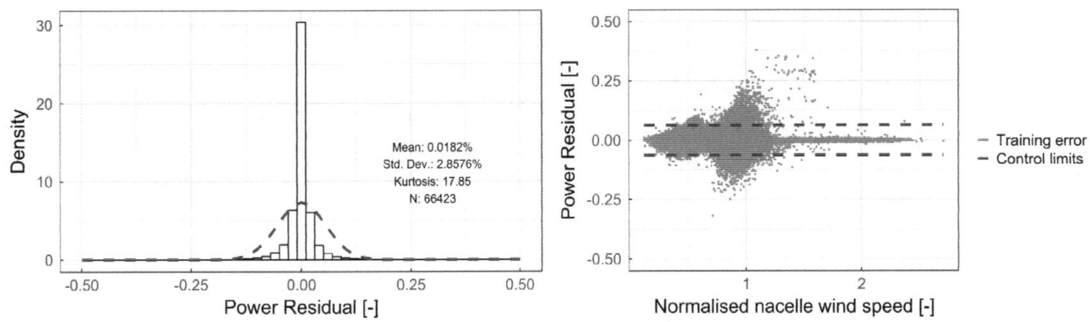

Figure 7. Training error statistics and distribution from RF model (left side); the related normal distribution is illustrated in blue. Training error (grey) and related control limits (blue) from RF model varying depending on the wind speed (right side).

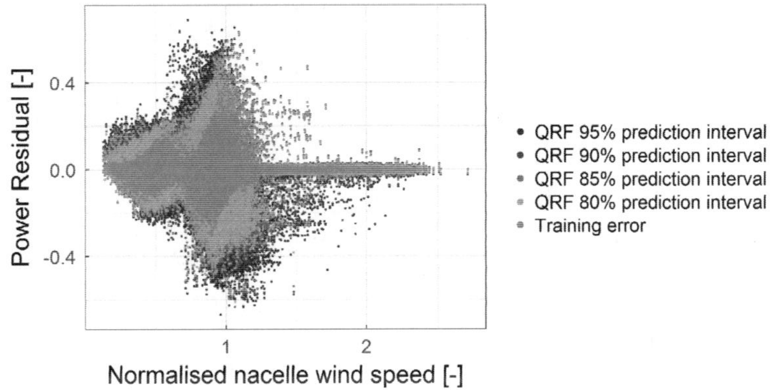

Figure 8. Training error (grey) and related control limits (4 shades of blue) from the QRF varying depending on the wind speed.

3.4. Time resolution detection capabilities

The detection capabilities of each model were tested using two different data time resolutions. To this end, the first month of the year was used to train the models and the subsequent months were used to generate operating residuals. For every model, an underperformance event was deemed to be every time the actual power was below the related low performance limit.

The control limits for normal performance modelled by the kNN and the RF were obtained based on Equation 1 and the following. First, η was obtained from a theoretical definition of the desired level of confidence (95%) given the assumed normal distribution, so-called *Normal error*. Second, η was adjusted empirically so that the resulting control limits encompassed 95% of the training observations, so-called *Adjusted error*.

The cumulative daily number of underperformance events detected was considered for WT performance monitoring purposes. The results for one of the healthy turbines using the high-frequency SCADA data and the different modelling approaches are shown in Figure 9. Similarly, the results using 10-minute data are shown in Figure 10.

IOP Conf. Series: Journal of Physics: Conf. Series **926** (2017) 012009 doi:10.1088/1742-6596/926/1/012009

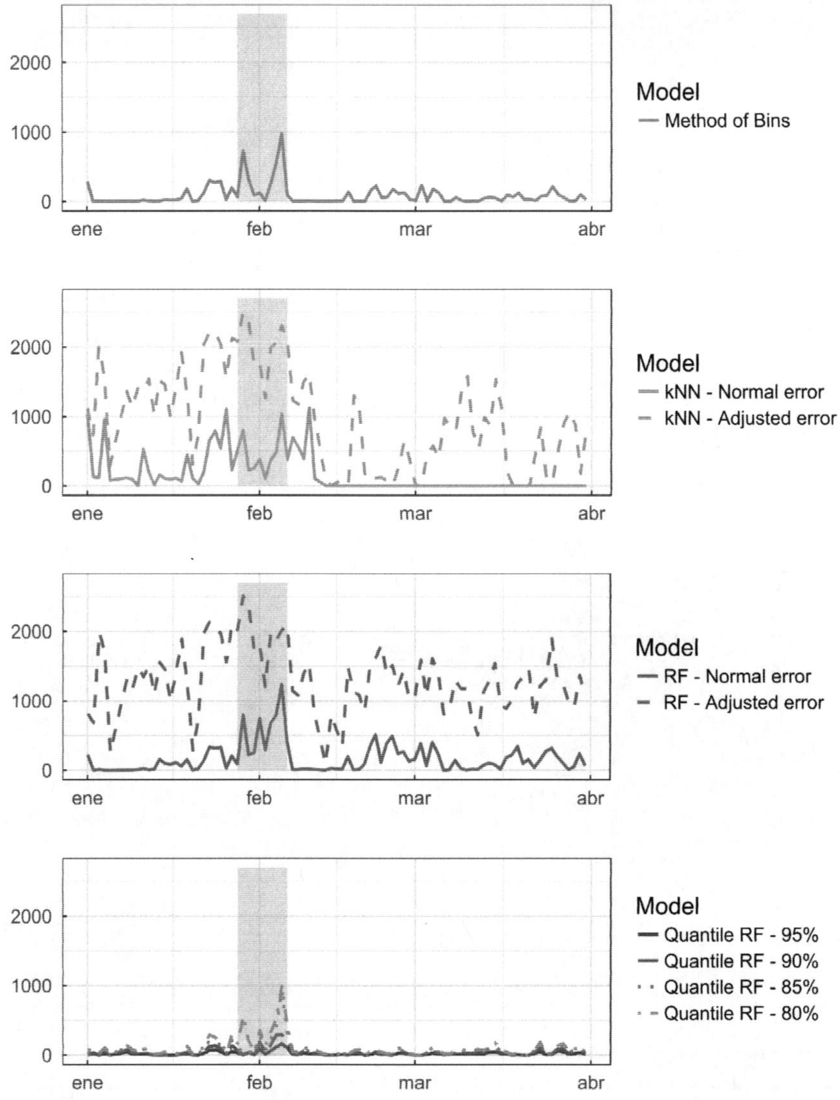

Figure 9. Daily monitoring of a healthy WT using high-frequency SCADA data: cumulative daily number of underperformance events detected with four different modelling approaches. The blue shaded area represents a storm event.

As reported in the alarm logs, a storm occurred in the beginning of February, highlighted in blue in both Figures 9 and 10. When using SCADA data of high-frequency, this event was effectively detected by the MOB and the QRF with confidence levels below 85%, while the latter showed the lowest level of false positives. The use of fixed control limits for the kNN and the RF, with the different uncertainty assessments, gives however a high level of false positives. Therefore, WT performance is not properly monitored when using these two approaches.

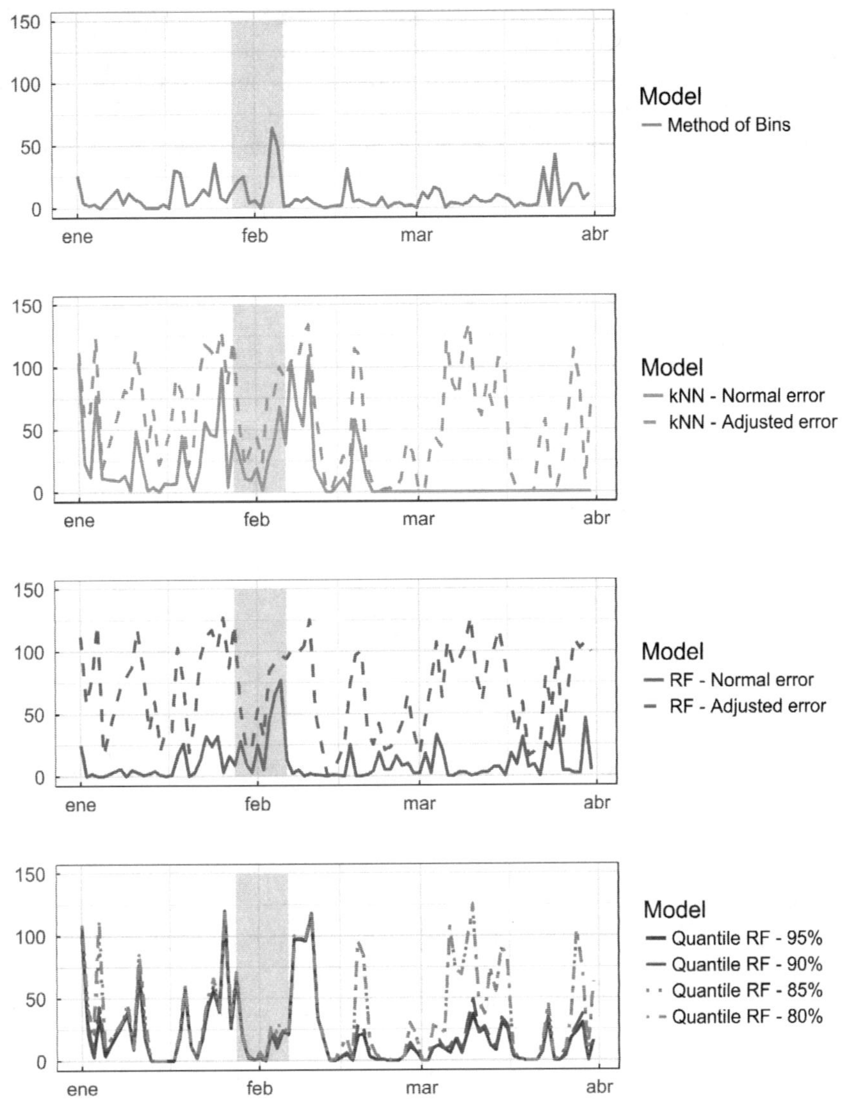

Figure 10. Daily monitoring of a healthy WT using 10-minute SCADA data: cumulative daily number of underperformance events detected with four different modelling approaches. The blue shaded area represents a storm event.

Similar results are observed in Figure 10 for *k*NN and the RF approaches using 10-minute data. Nevertheless, no significant deviation is detected by the QRF in Figure 10. The abnormal behaviour previously recognised is obscured with many irrelevant events, hindering the appearance of clear patterns of underperformance. As concerns the MOB, the deviation previously detected when using SCADA data of high-frequency is not as significant as when 10-minute data are used. Similarly, no clear patterns are identified. Summarising, WT performance was more accurately monitored in this case study by using high-frequency SCADA data together with QRF for performance assessment.

This storm event reported in the SCADA alarm logs was further investigated in terms of wind speed behaviour as illustrated in Figure 11. As can be seen, the wind speed recorded was not always higher than the cut-out during the whole event. As a result, the underperformance events successfully detected by the QRF using high-frequency SCADA data in Figure 9 were not all due to a normal stoppage occurring for wind speeds beyond the cut-out. Rather, these could also be related, for instance, to high turbulence intensity, and hence extreme loads, to high variations in wind direction or to high wind hysteresis effects.

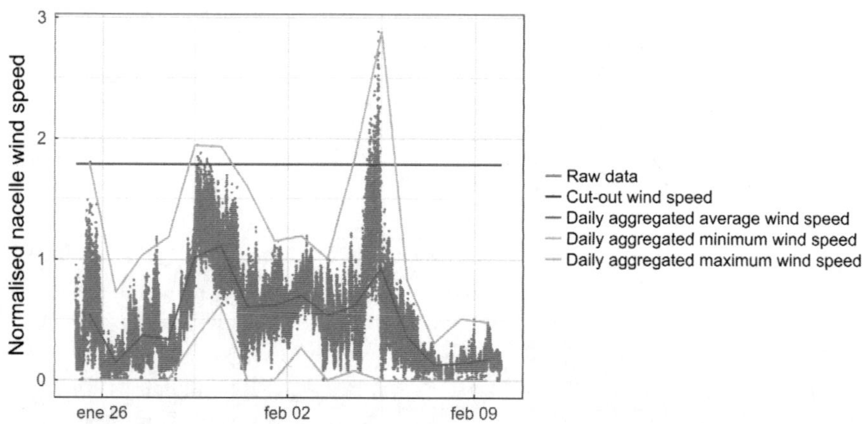

Figure 11. Nacelle wind speed data recorded during the storm event as reported in the alarm logs.

3.5. WT performance monitoring

High-frequency data and the QRF suggested approach were used to monitor the performance of the three turbines described in Section 3.1. WT performance was monitored during the three months prior to the occurrence of a major gearbox assembly failure. As detailed in Section 2, the alarms registered by the SCADA system were translated into component-related information and monitored concurrently to WT performance, to address the root-cause of the detected deviations.

The cumulative daily detected events of underperformance are illustrated in Figure 12 for the the three turbines, together with the alarms recorded at the same time by the SCADA system. The storm event detected in Figure 9 for WT2 is highlighted in blue. The methodology achieves its detection for the three turbines. Moreover, most of the alarms registered during this event were categorised as weather related events. It can be said that this extreme weather event was successfully captured as WT underperformance by the methodology.

Furthermore, another significant deviation was found by the methodology by the end of February in WT3. This event is also highlighted in red in Figure 12. A significant number of alarms were recorded by the SCADA system at the same time, all related to gearbox malfunction. As a result, it is very likely that this significant event of underperformance was due to an abnormal condition of the gearbox assembly. This significant event of underperformance, apparently due to an abnormal condition of the gearbox assembly, was detected 34 days before the actual appearance of the gearbox failure. A potential explanation for this performance deviation detected could be the occurrence of a torque reversal due to the transient loads caused by the developing failure of the gearbox assembly. This would also explain the transitional nature of the detected event. Nevertheless, this could not be confirmed with certainty due to the lack of data.

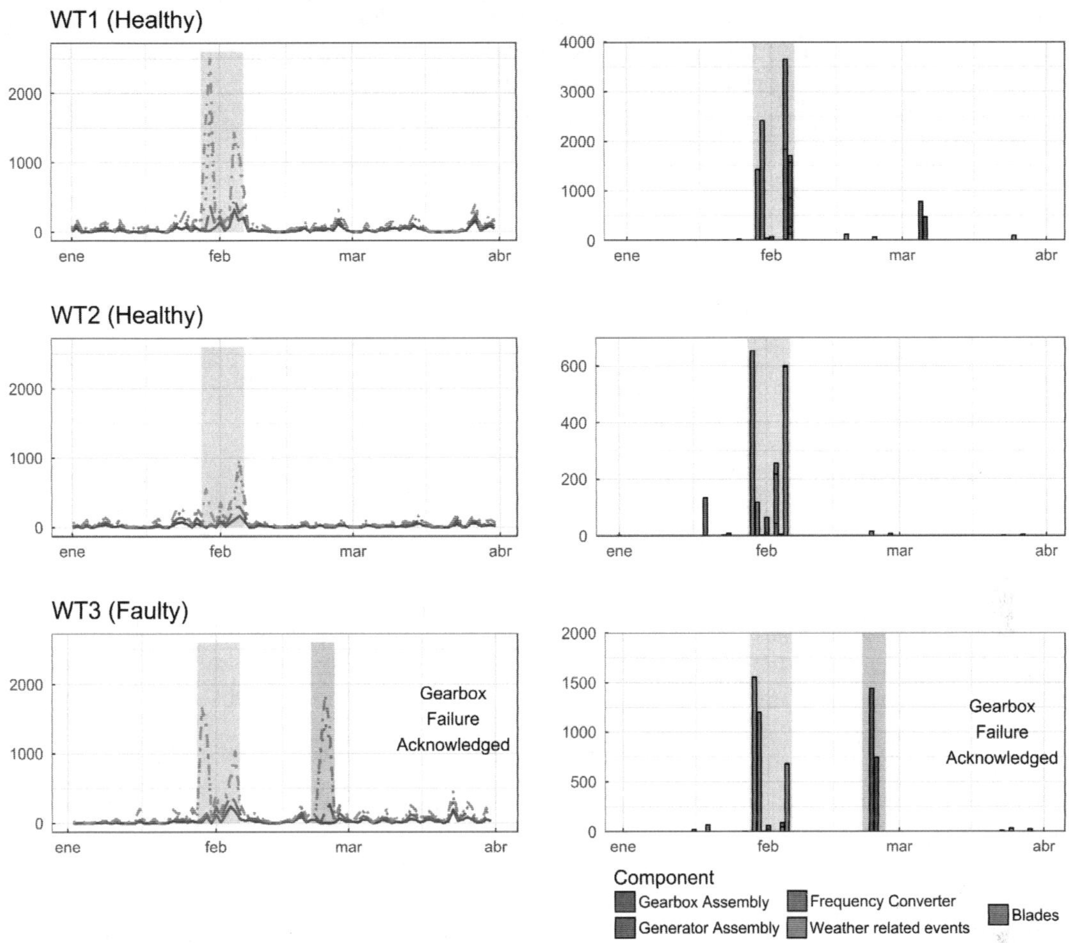

Figure 12. Detection of daily underperformance events (left) and corroborated component-related root cause (right). Alignment of the two are noted by their co-occurence in the shaded regions; the blue shaded area corresponds to the storm event and the red shaded area corresponds to the underperformance detected in the faulty WT.

In any case, the methodology appears to be successful in the monitoring of WT performance, in the detection of performance deviations and in the assessment of their related root cause in the presented case study. The transitional nature of the detected events confirms the importance of using high-frequency data.

4. Conclusions and outlook

This paper presents a methodology where high-frequency SCADA data and QRF are used to monitor WT performance and ascertain the root-cause in the event of underperformance. The results confirm the effectiveness of the methodology and its usefulness as a condition monitoring tool through several case studies.

Monitoring performance is fundamental during the O&M phase, to ensure every WT in a WF is operating safely and efficiently. However, modelling a reference performance together with its related uncertainty is still challenging. Multivariate non-parametric models are confirmed in this paper as providing accurate representation of WT performance. Moreover, the use of QRF is first introduced for a probabilistic assessment of WT performance. This empirical approach allows the model to take into account the heteroscedastic nature of the model residuals and introduces the concept of probabilistic assessment of WT performance.

The use of high-frequency data provides a deeper understanding of WT performance and its natural variability. As illustrated in several case studies it improves the detection capabilities of the suggested approach.

Moreover, the integration of the alarm logs to ascertain the root-cause for the underperformance events detected may turn the suggested approach into a high-level condition monitoring tool with great potential of applicability. Indeed, the presented methodology may be particularly useful for very large WFs, as found offshore. This tool could rapidly identify underperforming WTs that may be in need of a more in-depth analysis. Furthermore, the suggested approach does not require any additional devices and investments, such as LIDAR or meteorological masts.

Further testing and evaluation are needed to confirm the detection capabilities of the suggested methodology; that is, in the case of failures of other components, or different failure modes of the same component. Indeed, questions need to be asked about the relation between WT abnormal performance and specific failure modes. Unfortunately, this investigation is subject to real data availability. Current research is also performed to compare the effectiveness of the suggested approach with other methods for WT performance probabilistic assessment.

Finally, few operators are aware of the potential for SCADA data to reinforce CMS because of its low sampling rate. WF operators are therefore highly encouraged to store high-frequency data to build effective predictive maintenance strategies in order to optimise the O&M phase.

Acknowledgments

This project has received funding from the European Union's Horizon 2020 research and innovation programme under the Marie Sklodowska-Curie grant agreement No 642108.

The authors would like to thank Enel Green Power for their cooperation and supply of the datasets. The authors gratefully acknowledge Tim Rubert (University of Strathclyde) for his helpful suggestions and comments. The anonymous reviewers are also acknowledged for their comments.

References

[1] Carroll J, McDonald A, Dinwoodie I, McMillan D, Revie M and Lazakis I 2017 *Wind Energy* **20** 361–378
[2] García Márquez F P, Tobias A M, Pinar Pérez J M and Papaelias M 2012 *Renewable Energy* **46** 169–178
[3] Yang W, Tavner P J, Crabtree C J, Feng Y and Qiu Y 2014 *Wind Energy* **17** 673–693
[4] McMillan D, Thöns S and May A 2015 *IET Renewable Power Generation* **9** 900–907
[5] Tautz-Weinert J and Watson S J 2017 *IET Renewable Power Generation* **11** 382–394 ISSN 1752-1416
[6] Kusiak A, Zhang Z and Verma A 2013 *Energy* **60** 1–12
[7] Cambron P, Lepvrier R, Masson C, Tahan a and Pelletier F 2016 *Renewable Energy* **94** 126–135
[8] Gill S, Stephen B and Galloway S 2012 *IEEE Transactions on Sustainable Energy* **3** 94–101
[9] Kusiak A, Zheng H and Song Z 2009 *Renewable Energy* **34** 1487–1493
[10] Uluyol O, Parthasarathy G, Foslien W and Kim K 2011 *Annual Conference of the Prognostics and Health Management Society* August pp 1–8
[11] Lydia M, Selvakumar A I, Kumar S S and Kumar G E P 2013 *IEEE Transactions on Sustainable Energy* **4** 827–835
[12] Taslimi-Renani E, Modiri-Delshad M, Elias M F M and Rahim N A 2016 *Applied Energy* **177** 544–552
[13] Schlechtingen M, Santos I F and Achiche S 2013 *IEEE Transactions on Sustainable Energy* **4** 671–679
[14] Karlsson D 2014 *Wind Turbine Performance Monitoring using Artificial Neural Networks With a Multi-Dimensional Data Filtering Approach* Master's thesis Chalmers University of Technology
[15] Jin Zhou, Peng Guo and Xue-Ru Wang 2014 *2014 International Conference on Machine Learning and Cybernetics* (IEEE) pp 71–76
[16] Janssens O, Noppe N, Devriendt C, de Walle R V and Hoecke S V 2016 *Engineering Applications of Artificial Intelligence* **55** 331–338

[17] Wilkinson M 2016 *Technical Workshop on Analysis of Operating Wind Farms* ed WindEurope (Bilbao: WindEurope)
[18] Gonzalez E and Melero J J 2017 *30th International Congress & Exhibition on Condition Monitoring and Diagnostic Engineering, COMADEM*
[19] Zhang J, Jain R and Hodge B M 2016 *Energy* **112** 1139–1152
[20] St Martin C M, Lundquist J K, Clifton A, Poulos G S and Schreck S J *Wind Energy Science* 221–236
[21] Wagner R, Antoniou I, Pedersen S M, Courtney M S and Jørgensen H E 2009 *Wind Energy* **12** 348–362
[22] Kusiak A and Verma A 2013 *IEEE Transactions on Sustainable Energy* **4** 192–199
[23] Koenker R and Hallock K F 2001 *Journal of Economic Perspectives* **15** 143–156
[24] Meinshausen N 2006 *The Journal of Machine Learning Research* **7** 983–999
[25] Reder M, Gonzalez E and Melero J J 2016 *Journal of Physics: Conference Series* **753** 72027
[26] Gonzalez E, Reder M and Melero J J 2016 *Journal of Physics: Conference Series* **753** 72019
[27] International Eletrotechnical Comission 2005 IEC 61400-12-1 Wind turbines Part 12-1: Power performance of electricity-producing wind turbines. Standard
[28] International Eletrotechnical Comission 2013 IEC 61400-12-2 Wind turbines Part 12-2: Power performance of electricity-producing wind turbines based on nacelle anemometry. Standard
[29] Witten I H and Frank E 2005 *Data Mining: Practical Machine Learning Tools and Techniques* 2nd ed (San Francisco, CA: Morgan Kaufmann)
[30] Breiman L 2001 *Random Forests* vol 45 (Springer)
[31] Mennitt D, Sherrill K and Fristrup K 2014 *The Journal of the Acoustical Society of America* **135** 2746–2764

Investigation of the relationship between main-bearing loads and wind field characteristics

E Hart[1], A Turnbull[1], D McMillan[1], J Feuchtwang[1], E Golysheva[2] and R Elliott[2]

[1] Wind and Marine Energy Systems CDT, University of Strathclyde, Glasgow, UK
[2] Romax InSight, University of Nottingham Innovation Park, Nottingham, UK

E-mail: edward.hart@strath.ac.uk

Abstract. This paper investigates the relationship between main bearing loads and the characteristics of the incident wind field in which a wind turbine is operating. For a 2MW wind turbine model, fully aeroelastic multibody simulations are performed in 3D turbulent wind fields across the wind turbines operational envelope. Hub loads are extracted and then injected into a simplified drivetrain model of a single main-bearing configuration whose parameters are determined using finite element software. The main bearing reaction loads and load ratios from the simplified model are presented and analysed. The results indicate that there is a strong link between wind field characteristics and the loading experienced by a single main-bearing, with more damaging load ratios seen to occur in low turbulence and high shear wind conditions.

1. Introduction

Wind turbine main-bearings seldom reach their design lives of roughly 20 years, with many failing in less than six [1]. The reasons for this are multi-faceted and as yet not fully understood; certainly the conditions experienced by the main-bearing inside a wind turbine drivetrain deviate significantly from the well understood and relatively constant conditions present in conventional energy plant generators. It is in these latter and altogether more hospitable conditions where much of the current main-bearing knowledge base has been developed and, when dealing with wind turbines, there is generally an implicit assumption that this prior experience can be reapplied without much alteration. The relatively high proportions of main bearing failures seen in the field suggest that these assumptions need to be revisited. This paper seeks to both challenge these assumptions and attempt to shed some light on the problem by addressing the following question:

What range of load conditions are experienced by a wind turbine main-bearing across its operating envelope and to what extent can these load conditions be linked to characteristics of the incident wind field?

IOP Conf. Series: Journal of Physics: Conf. Series **926** (2017) 012010 doi:10.1088/1742-6596/926/1/012010

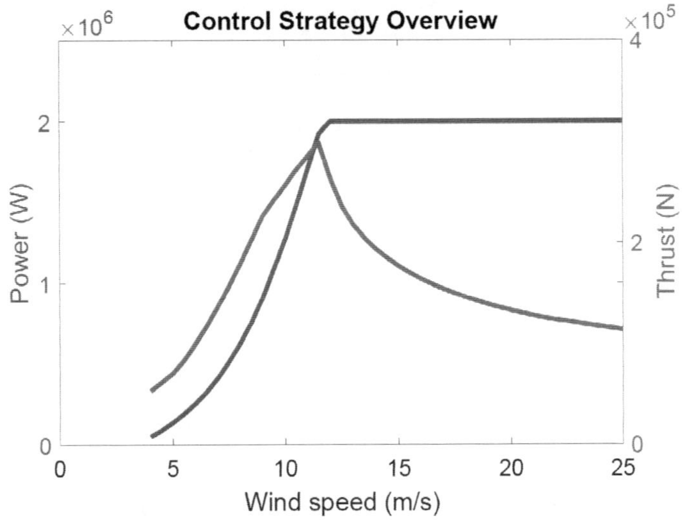

Figure 1. Design power and thrust curves for the simulated wind turbine.

2. Methodology

The current paper focuses on the case of a single main-bearing. The two main-bearing case will be considered in future work.

The investigations proceeded in three distinct stages:

(i) Generation of 3D turbulent wind fields and fully aeroelastic simulations of complete wind turbine system

(ii) Extraction of hub loads from simulations and injection into simplified drivetrain model

(iii) Analysis of main-bearing reaction forces obtained from simplified drivetrain model.

Fully aeroelastic multi-body simulations of a 2MW wind turbine were performed in GH Bladed software and the 3D wind fields generated in compliance with IEC standards [4] using a Kaimal spectrum. The wind fields used in this case were generated as in [5] where the relevant theory is also discussed. As a brief summary: the second order wind field statistics are captured in frequency space by what is known as a spectral tensor, a function of wavenumer whose values are related to the energy in the wind. For a given spectral tensor model (the Kaimal spectrum being one) velocity components can be determined via summing a set of coefficients, determined for that wind model, each multiplied by unit variance random numbers. The presence of random numbers means that, using different initial random number seeds, it is possible to generate wind fields which are different, but share the same second order statistics. Summary data for the wind turbine model is given in Table 1 and its design power and thrust curves are given in Figure 1. Figure 2 shows a schematic of the drivetrain model, including the reference frame used. Forces along a given axis, i say, will be denoted by B_i for input forces and F_i for bearing reaction forces, and similarly moments will be denoted by M_i. In order to obtain an understanding of the loads experienced by the main bearing, we treat the main bearing as a simple support. The hub loads are then decomposed into their orthogonal components and in-plane force-moment pairs injected into models of the form shown in Figure 3. For example, M_z and B_y will be injected into the same model. The bearing is assumed to react all of the thrust forces. The length L_1 is the distance (supplied by Romax) from the hub to the main-bearing on a commercially available wind turbine of the same size as the simulated model. The length L_2 is determined by Romax finite element software to be the equivalent length of the torque arm from the main-bearing to

Table 1. Data for simulated wind turbine.

Rated power	2MW
Rotor diameter	80m
Blade number	3
Hub height	61.5m
Aerodynamic control	Pitch
Fixed/Variable speed	Variable
Controller	PI

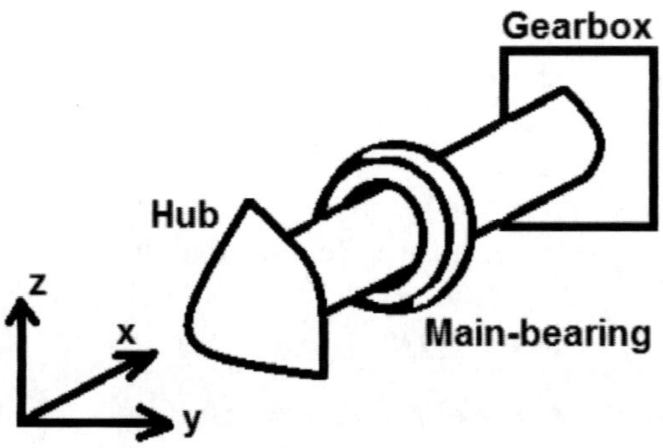

Figure 2. Drivetrain configuration, including reference frame.

the gearbox. Gearbox spring stiffnesses are also available, however, the current analysis has been performed in static equilibrium and since the system is determinate, these have not yet been required. The outputs from these simplified models is then re-combined, giving approximations of the resultant reaction forces on the main bearing. The main outputs investigated in this work are the axial to radial load ratio (F_x/F_R where $F_R = \sqrt{F_y^2 + F_z^2}$) and peak loading in both the axial and radial planes. The load ratio has been found to be important for bearing health, since high values of F_x/F_R can lead to damage modes related to skewing, sliding and the unseating of the bearing row [2, 3].

Simulations were performed in GH Bladed software for the wind turbine model outlined above. IEC design requirements were followed throughout the simulations [4]. The process and analysis followed here closely resembles that of [5]. 6 different 3D turbulent wind fields were generated for each combination of: mean wind speed (10, 12, 16, 20 m/s), turbulence intensity (TI) (low, med and high as specified by the IEC [4]) and power law shear exponent (0.2, 0.6); a total of 144 wind fields. The turbine was simulated in each of the generated wind fields and the hub loading, which includes gravitational and inertial forces, was extracted. Each of these simulations resulted in 10 minutes of hub loading time history.

3. Results

Figure 4 show a histogram of the axial to radial bearing reaction load ratios, F_x/F_R, for 10m/s mean wind speeds, medium turbulence intensity and both low and high shear exponents. It

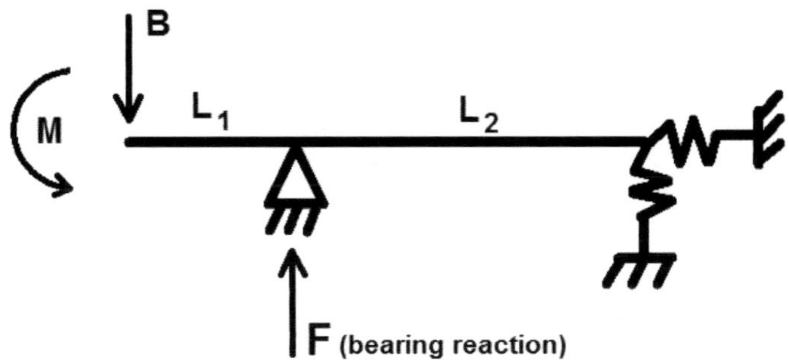

Figure 3. Simplified drivetrain model for each in-plane force-moment pair.

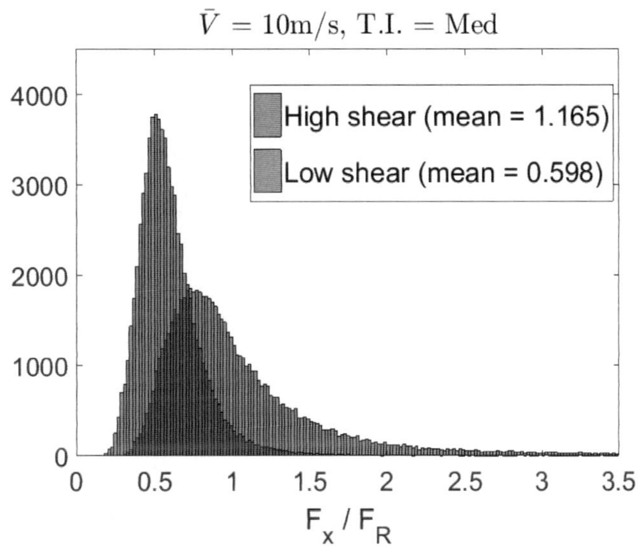

Figure 4. Histogram of axial to radial bearing reaction force ratios across simulations with a mean wind speed of 10m/s, medium turbulence intensity and low (0.2) and high (0.6) shear exponents.

can be seen that the load ratios have a distribution which is Weibull in nature. An increase in shear profile has a significant effect, resulting in a marked shift of the mean towards higher load ratio values. It was found that all of these plots at various mean wind speeds and turbulence intensities were qualitatively similar, hence, it is the position of the mean values which is of most interest in order to make comparisons. Figure 5 shows the values of these mean main-bearing load ratios across the whole simulation range. There are two very clear results which can be seen in this plot. The first being that shear exponent is the single most important factor in determining the mean load ratio for the main bearing reaction forces. The second is that there is noticeably different reaction behaviour between regions where pitch control is active and regions where it is not, the transition between these two regimes is captured by the high shear and high turbulence intensity point with 12 m/s mean wind speed. Furthermore, in general the mean load ratios experienced by the main-bearing are not sensitive to the turbulence intensity, except in the pitch active regions (16 and 20 m/s) and with the high shear exponent. It is of particular note that for the above rated operational region, where pitch is active, the highest mean load

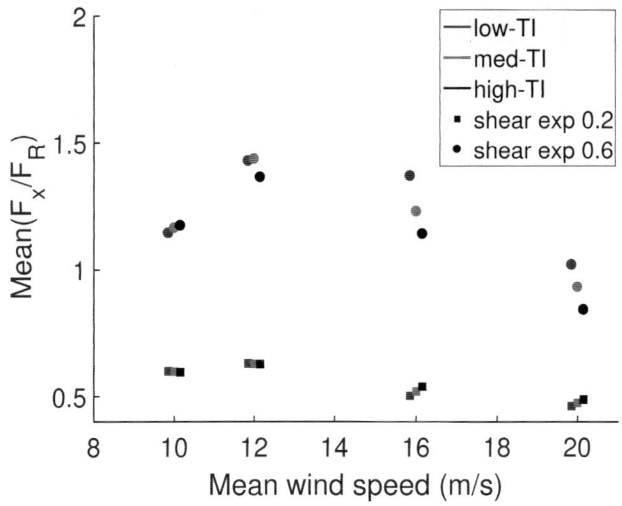

Figure 5. Mean axial to radial load ratios. Note, results have been staggered about each mean wind speed for clarity.

ratios (and so potentially most damaging in terms of bearing misalignment) are experienced in high shear and low turbulence intensity conditions.

Peak reaction loads on the main-bearing in both the axial and radial directions are also important drivers of mainbearing lifetime. In order to investigate these, peak axial and radial loads were found from each main-bearing reaction time history generated using the drive-train model. The means and standard deviations of peak loads across the six wind fields for each wind field type are shown in Figures 6 and 7 for radial and axial loads respectively. Peak radial loads can be seen to generally increase steadily with mean wind speed and the main driver for changes is again the shear exponent. It is important to note that increasing the shear exponent results in a decrease in the peak radial loads, this make intuitive sense since increased imbalance in the rotor is acting contrary to gravitational forces, thus canceling out some loading. The peak radial loads are clustered fairly tightly and have small enough standard deviation, especially at lower wind speeds, to imply that the loads are strongly linked to wind field characteristics.

Peak axial loads on the other hand generally decrease with mean wind speed, as should be expected from the thrust curve in Figure 1, and the association to shear exponent is much less marked. Instead, it can be seen here that the main driver for peak axial loading is the turbulence intensity. Again it can be seen that the loads experienced by the main bearing are closely linked to the wind conditions in which the wind turbine is operating. Finally it can be seen that the results in Figure 5 cannot be inferred from the peak loading trends seen in Figures 6 and 7, since ratios here remain more or less in the same proportions, and this implies that there is some other coupling between these two loads which is dynamic in nature and cannot be reconstructed from extreme values alone.

4. Conclusions

The results seen here imply the following:

(i) There is a strong link between the forces experienced by a single main-bearing and the wind conditions in which the wind turbine is operating

(ii) Load ratios on the main bearing behave very differently in above and below rated operation,

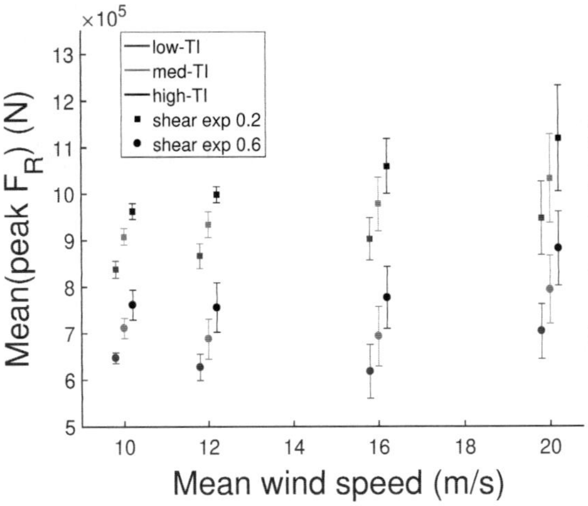

Figure 6. Mean peak radial loading over each set of six wind fields with parameters in common. Note, results have been staggered about each mean wind speed for clarity.

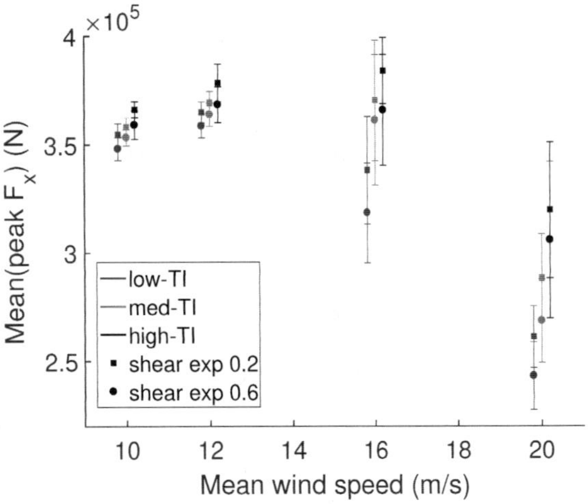

Figure 7. Mean peak axial loading over each set of six wind fields with parameters in common. Note, results have been staggered about each mean wind speed for clarity.

with the most damaging conditions occuring in low turbulence intensity and high shear across most wind speeds

(iii) Load ratios on the main bearing are most strongly driven by shear exponent in general, this is also the case for peak radial loading

(iv) Peak axial loading at a given mean wind speed is most strongly driven by turbulence intensity.

The results documented here also demonstrate the very variable nature of main bearing loads across the operating envelope of a wind turbine. This challenges the assumption that knowledge from more conventional power generation technologies can be reapplied in this case with little or no alteration. The result of doing this might well be contributing to main bearing failures.

There was seen to be strong links between the loads experienced by a single-main bearing and the characteristics of the incident wind field. This implies that effective health monitoring techniques for main-bearings may necessarily include wind field measurements and analysis.

Acknowledgements
This work has been funded by the EPSRC, project reference number EP/L016680/1.

References
[1] Sethuraman L et al 2015 *Main Bearing Dynamics in Three-Point Suspension Drivetrains for Wind Turbines* NREL, AWEA WINDPOWER, Orlando

[2] Meadows B and Shapiro J 2016 *Data-driven Main Bearing Maintenance Strategies to Reduce Unplanned Maintenance Costs* Wind Systems, inFOCUS: WINDPOWER

[3] Baldwin B *Increasing Bearing Reliability In A Main Shaft Support System* Timken, White Paper

[4] IEC 2005 *Wind Turbine Design Requirements* IEC 61400-1:2005(E)

[5] Hart E, Keegan M and McMillan D 2016 *A Lookup Table Approach to Determining Wind Turbine Operational Fatigue Loading from Wind Field Measurements* ASRANet, CORE Conference, Glasgow

IOP Conf. Series: Journal of Physics: Conf. Series **926** (2017) 012011 doi:10.1088/1742-6596/926/1/012011

Automatically Identifying and Predicting Unplanned Wind Turbine Stoppages Using SCADA and Alarms System Data: Case Study and Results

Kevin Leahy[1], **Colm Gallagher**[1], **Ken Bruton**[2], **Peter O'Donovan**[1] **and Dominic T.J. O'Sullivan**[1]

[1]Department of Civil & Environmental Engineering, University College Cork, Ireland
[2]Department of Mechanical, Biomedical and Manufacturing Engineering, Cork Institute of Technology, Cork, Ireland

E-mail: kevin.leahy@umail.ucc.ie

Abstract. Using 10-minute wind turbine SCADA data for fault prediction offers an attractive way of gaining additional prognostic capabilities without needing to invest in extra hardware. To use these data-driven methods effectively, the historical SCADA data must be labelled with the periods when the turbine was in faulty operation as well the sub-system the fault was attributed to. Manually identifying faults using maintenance logs can be effective, but is also highly time consuming and tedious due to the disparate nature of these logs across manufacturers, operators and even individual maintenance events. Turbine alarm systems can help to identify these periods, but the sheer volume of alarms and false positives generated makes analysing them on an individual basis ineffective. In this work, we present a new method for automatically identifying historical stoppages on the turbine using SCADA and alarms data. Each stoppage is associated with either a fault in one of the turbine's sub-systems, a routine maintenance activity, a grid-related event or a number of other categories. This is then checked against maintenance logs for accuracy and the labelled data fed into a classifier for predicting when these stoppages will occur. Results show that the automated labelling process correctly identifies each type of stoppage, and can be effectively used for SCADA-based prediction of turbine faults.

1. Introduction

Due to the nature of the dynamic conditions in which they operate, operations and maintenance (O&M) costs for wind turbines can account for up to 30% of the cost of generation of energy [1, 2, 3]. In the case of offshore turbines, access for routine maintenance can be expensive and weather dependent, with the costs for unscheduled repairs or inspections being even higher. Transport costs for a maintenance crew to an offshore turbine ranges from €85/hr on relatively small vessels, to over €400/hr for larger vessels and helicopters which are not as dependent on calm sea conditions [4]. Because of this, it is important to have robust control and monitoring strategies which can detect faults and notify technicians without raising false alarms.

If a specific type of fault, along with an estimate for the time to failure, can be predicted in advance, preventative measures can be taken to either avoid the fault or schedule maintenance at the optimal time. For example, scheduling maintenance during good weather for offshore turbines, or during other planned maintenance activities for maximum economic benefit [4]. As well as this, information about any predicted turbine downtime that may be unavoidable is very useful to grid operators. In many jurisdictions, electricity markets have moved or are planning to move to a model which requires non-dispatchable generators to give a forecast for their daily energy generation. In Ireland, the new Integrated Single Electricity Market will have this requirement from 2018 [5].

Fault prognostic functions have traditionally been performed by condition monitoring systems (CMSs) based on vibration or oil particulate sensors or similar. These can cost upwards of €13,000 per turbine and, due to the inherent wide range of operating conditions associated with wind energy, have not performed as well as CMSs in other industries [6, 7]. CMSs leveraging pre-existing low-frequency (generally 10-minute period) SCADA data can avoid the high installation costs associated with traditional CMSs [8]. Using this data to to detect, diagnose and predict faults and turbine downtime has seen some success in the past. However, problems with data availability and some ambiguity in how faults are defined have complicated efforts.

1.1. Related Works

Evidence of performance degradation leading up to large or catastrophic faults has been well documented in the literature. In [9], the authors showed that normal behaviour models showed a residual that indicated a fault was imminent leading up to major maintenance works. The authors in [10] also use a residual model for prognostics on the main bearing of a wind turbine. Their work shows that some indication of a fault is possible more than 30 days in advance of failure, but that further testing is needed to verify. The work done in [11] once again uses normal behaviour modelling, this time to detect anomalous temperatures indicating generator and gearbox faults. The results show that some indications of faults are possible in advance, but again more work is needed to prove effectiveness.

Other work has shown varying degrees of success in predicting less serious, but more frequent faults, with shorter prognostic horizons. These methods generally use a classification-based approach, where SCADA samples are classified as "possibly faulty" or "fault-free" in advance of a fault occurring. The authors in [12] found that prediction of a specific type of fault was possible with 67% accuracy and 73% recall 30 minutes in advance of the fault occurring. However, here, the test set used for evaluation had undersampled the fault-free data and did not represent a true distribution. In [13], the authors use the RIPPER decision tree algorithm to derive human-readable rules for giving an indication of pitch fault occurrence, with precision and recall scores of 0.8 and 0.75, respectively, on a 48 hour prognostic horizon. Once again, however, these scores were based on a validation set which undersampled the fault-free class. When tested against a full set of data, recall scores were still around 87%, but precision was reduced to 25%. However, there was a 52% reduction in the number of false alarms when compared to the SCADA alarm system, lowering the burden of analysis for maintenance operators. The prediction of specific blade pitch faults was demonstrated in [14] using genetically programmed decision trees. Here, the maximum prediction time was 10 minutes, at a 68% accuracy and 71% recall. However, the SCADA data used was at a resolution of 1s rather than the more common 10 minutes. Previous work by the authors in [15] showed that prediction of generator heating faults was possible one hour in advance with a recall score of >0.9. This performance was brought down by poor precision of 0.56, meaning there were too many false alarms for it to be field deployable, though this score was significantly improved with the construction of additional features in [16].

1.2. Defining a Fault

In all of the previous work mentioned, a significant point to keep in mind is how "faults" are defined. In some cases, such as [12, 15], periods when turbine "fault alarms" were active are used to indicate the presence of a fault. These are alarms generated by the turbine SCADA system and are generally triggered when the controller detects an operating condition outside of acceptable bounds, e.g. over-speed, over-temperature or a control issue. Hence, single instances

IOP Conf. Series: Journal of Physics: Conf. Series **926** (2017) 012011 doi:10.1088/1742-6596/926/1/012011

of turbine alarms are not always indicative of a fault having taken place, and can often be false alarms [17]. In other cases, faults are defined as being the result of a maintenance action taking place, as depicted in the maintenance logs [13]. This avoids the issue of clouding the data with false alarms and very brief stoppages intended to reset the control system rather than as the result of damage or to avoid damage taking place. However, labelling data using maintenance logs can be a tedious process, as logs are stored as written documents rather than in dedicated databases [18]. As well as this, the logs are not always standardised across different OEMs or operators, or, in some cases, even different maintenance events, meaning automating this process by using OCR text recognition or reading spreadsheets can be very involved.

One way to get the best of both worlds is by identifying periods of downtime and only labelling periods where downtime was significant (e.g. >12 hours) as being indicative of a fault taking place. Although this avoids the problems associated with false alarms, it means that some shorter term stoppages are not predicted, and the root cause or even sub-system which the fault belongs to are not immediately clear.

In this paper we present a method for identifying periods of faulty operation in the SCADA data without needing to manually reference maintenance logs by assigning different rules to generated alarms, and use this to perform fault prediction using classification techniques. In section 2, we give an overview of the data used in this study. In section 3, we describe the automated fault and stoppage labelling process and evaluate its accuracy against maintenance logs and availability data. In section 4, the classification techniques used to predict when these faults occur are discussed. In section 5, the results of this fault prediction are discussed, while conclusions are drawn and future work described in section 6.

2. Description of Data

The data used in this study composes of 6 months of data from 11 2.5MW DFIG wind turbines in the East of Ireland. The dataset comprises of three parts: SCADA data, alarms data and availability data. Additionally, maintenance logs, which consist of spreadsheet and PDF documents, are used to validate the automated fault-identification process.

2.1. SCADA Data

The turbine control system monitors many instantaneous variables such as power output, wind speed, temperatures of various components, blade angles, etc. These are aggregated into 10 minute averages, and in some cases the maximum, minimum and standard deviation of values that occurred in each 10 minute period are also included. In this case, there were over 300,000 samples spread across all turbines, with each sample having over 90 different variables. A sample of this data is seen in table 1.

Table 1. Ten Minute SCADA Data

TimeStamp	Wind Speed (avg.) m/s	Wind Speed (std.) m/s	Power (avg.) kW	Gen. RPM (avg.) rpm	Bearing Temp (avg.) °C
09/06/2015 14:10:00	5.8	0.1	367	756	25
09/06/2015 14:20:00	5.7	0.1	378	750	25
09/06/2015 14:30:00	5.6	0.5	384	747	25
09/06/2015 14:40:00	2.8	0.9	0	0	24

Table 2. Random Sample of Alarms Data from Turbine 2

t_s	t_e	Code	Description	Category	Severity
2015/06/01 02:13:38	17:25:05	a_{41}	Normal Operation	No Fault	Information
2015/06/02 10:19:05	10:19:05	a_{124}	Motor Fuse	Generator	Warning
2015/06/03 07:56:14	07:57:32	a_{91}	Low wind cut out	Weather	Information
2015/11/04 08:38:27	08:38:37	a_{81}	Line CCU current	Freq. Conv.	Fault
2016/02/01 01:26:46	01:26:54	a_{22}	Normal Operation	No Fault	Information
2016/02/02 15:05:38	15:11:01	a_{51}	Emergency stop	User	Critical Fault

2.2. Alarms Data

Turbine alarm systems vary between manufacturers, but generally behave in similar ways. At the most basic level, a turbine alarm (also called status message or event by some manufacturers) is generated to indicate that the turbine's operating state has changed. OEMs generally attribute at least three different levels of severity to turbine alarms:

- *Information alarms* are generally to communicate changes in certain operating conditions, e.g. when the wind speed is too low for generation, or a manual switch has been engaged.
- *Warning alarms*, on the other hand, are generated when the control system detects operating conditions or control variables that come close to exceeding certain thresholds.
- *Fault alarms* are generated when these thresholds are exceeded.

Fault alarms generally cause the turbine to shut down. Depending on the severity of the fault alarm itself, either an automatic remote reset, a manual remote reset, or a manual on-site inspection is required to reset the alarm. Alarms which require on-site inspections are generally alarms which do not often give false positives, i.e. they are usually related to safety critical components and only fire if these components are damaged or not operating correctly (e.g. the emergency braking system). In general, however, alarms which only need a remote reset to clear can often generate a lot of false positives. Due to the sheer volume of individual alarms generated, they can be overlooked by maintenance operators [17]. A sample of alarms data used in this study can be seen in table 2.

Altogether, there were over 100,000 alarm instances in the dataset. As can be seen, each instance has an associated start time, t_s, end time, t_e, code (here made anonymous for identity purposes), description, category and type. The category relates to its functional location on the turbine, while the severity refers to whether the alarm is an information, warning or fault alarm. Note that fault alarms for this particular manufacturer are also split into "fault" and "critical fault" categories, with "critical fault" alarms requiring a manual on-site reset to restart the turbine.

2.3. Availability Data

The turbine availability data tracks the availability states of the turbine over every 10 minute period in the SCADA data. These are essentially counters which document how many seconds in every 10 minute period the turbine was in one of the following states:

- SOT (System OK) - the turbine was operating normally
- DT (Down Time) - the turbine was down due to a fault on the turbine
- LOT (Line Out Time) - the turbine was not operating due to a grid fault
- WOT (Weather Out Time) - the turbine was not operating, or was curtailed, because of severe weather

- MT (Maintenance Time) - the turbine was down for routine maintenance
- RT (Repair Time) - the turbine was down for unplanned repairs

A sample of this data can be seen in table 3.

Table 3. Ten Minute Availability Data

TimeStamp	SOT	DT	LOT	WOT	MT	RT
09/06/2015 14:10:00	600	0	0	0	0	0
09/06/2015 14:20:00	400	200	0	0	0	0
09/06/2015 14:30:00	0	600	0	0	0	
09/06/2015 14:40:00	100	500	0	0	0	0
09/06/2015 14:50:00	600	0	0	0	0	0

3. Labelling Faults

As mentioned previously, labelling the SCADA data for fault classification can be a tedious step. Using maintenance logs involves a lot of manual intervention, while simply labelling the data according to times when alarms were active can mean that a lot of false positives are labelled as faults. As well as this, generally during a turbine stoppage related to a fault, there can be a number of alarms generated related to different subsystems, so it is not always clear which subsystem the original fault is attributed to. In this work, we outline a novel method of identifying the periods of faulty operation by cross referencing with particular sequences of alarms.

First, alarms which cause the turbine to shutdown, or only appear as a result of the turbine shutting down, are identified. We name this set of alarms A_s. These alarms are then assigned one of the following stop categories:

- Fault categories - These are alarms which cause the turbine to shut down due to a fault on the turbine. These are labelled according to the subsystem they belong to (note that these can differ from the OEM-assigned category seen in table 2):
  - *fc* - frequency converter faults
  - *pt* - pitch faults
  - *gb* - gearbox faults
  - *gn* - generator faults
  - *sn* - sensor faults
  - *to* - tower faults
  - *az* - azimuth/rotor faults
  - *bk* - brake faults
- *no* (normal operation) - the alarm causes the turbine to shut down, but due to things that are considered part of normal operation, e.g. cable untwisting, or running system tests.
- *wa* (weather) - the turbine is down due to severe wind conditions
- *ma* (manual/maintenance) - the alarm causes the turbine to shut down as a result of manual intervention, e.g. maintenance activities or a remote manual shut down.
- *gd* (grid) - these alarms notify the operator that the turbine has been shut down as a result of a grid fault, i.e. a fault not caused by the turbine

3.1. Alarm Batches

Any time the turbine shuts down, a number of alarms are generated. These alarms may persist and be active for the duration of the shutdown, or may be cleared before the turbine comes back online. The gap between the alarm being "cleared" (usually as a result of a remote reset) and the turbine coming back online and generating, however, can be a number of hours if the turbine needs a site visit for manual inspection. As well as this, many alarms may be triggered concurrently during a stoppage, meaning that it is difficult to attribute the stoppage to any single alarm. This means that if trying to identify periods of faulty operation to label SCADA data for fault prediction, labelling data according to specific instances of alarms and the times they were present can lead to erroneous classification.

To avoid this problem, we instead look at *batches* of alarm sequences that appear during individual turbine stoppages. The first step in creating these batches from the data is by identifying the alarm code that signifies the turbine returning to normal operation after a stop, referred to here as a_n. Each batch is then created for every turbine as follows (note that "A_s alarm" here refers to an alarm with a code from the set of alarms A_s):

(i) Find the earliest occurring alarm from set of alarms A_s in the data. Store its t_s as t_{start}.

(ii) Find the earliest occurring a_n alarm with $t_s > t_{start}$. Store its t_s as t_{end}

(iii) Find *all* alarms (i.e. not just including A_s alarms) which have start times that satisfy:

$$t_{start} \geq t_s < t_{end}$$

(iv) Select the next earliest occurring A_s alarm with $t_s > t_{end}$. Store its t_s as t_{start}.

(v) Repeat steps (ii)-(iv) until no more A_s alarms in the data.

In this way, any time the turbine was stopped, there was an associated batch with the set of alarms that appeared during this stoppage. A sample of a batch can be seen in table 4. Each alarm is listed according to its start time, the stop category of the alarm (note that some alarms are not A_s alarms, and so have no stop category), and the associated description. The alarm codes are left out for the sake of brevity.

As can be seen, the batch starts with two frequency converter fault related alarms at around 03:00 am. A number of other alarms are then generated in response to this, including pitch and

Table 4. Example of a batch of alarm sequences

t_s	Cat.	Description
24/12/2015 03:02:01	*fc*	Rotor conv. voltage fault
	fc	Invalid conv. response
24/12/2015 03:02:40	*pt*	Blade angle asymmetry
	pt	Pitch Motor Protection
24/12/2015 03:03:02	*pt*	Blade braking time high
24/12/2015 13:06:22	*ma*	Safety Chain Pulled
	az	Yaw motor over temp.
	pt	Pitch Comms Error
24/12/2015 13:06:57	*pt*	Pitch malfunction
24/12/2015 13:26:58	*pt*	Pitch System Test
24/12/2015 13:33:23	N/A	Idling Command
24/12/2015 13:46:16	N/A	Start-up command
24/12/2015 13:48:34	N/A	Spinning Up
24/12/2015 13:49:57	N/A	System OK

azimuth bearing/rotor alarms. These "triggered" alarms are not related to the initial fault, but are instead reactions to this initial fault. A detailed analysis of these types of alarm sequences can be found in [19]. The safety chain is then pulled the next day as the technician arrives on site to inspect the turbine. This also causes some additional alarms to trigger. A pitch system test is then performed (often, these periodic tests will automatically be performed when the turbine has been shut down to avoid needing to shut down especially for the test), before the turbine is given the command to start idling and come back online. The t_e for each alarm is not included, as all the alarms are cleared once the turbine comes back online. The total duration for this batch was just over 10 hours.

Each batch has a number of "root" alarms, which are the first alarms that appear when the turbine goes down. By looking at the stop category of these alarms, the entire batch can be attributed to being because of a fault in a particular subsystem, maintenance action, grid fault, etc. For example, in table 4, the stoppage can be attributed to the frequency converter as this is the stop category of the first two alarms in the batch. In some cases, where there are alarms with different stop categories in each set of root alarms, certain priority is given to some alarms. E.g. if a *ma* alarm appears along side some *pt* alarms, the stop is attributed to being because of a manual stop rather than a pitch fault. By associating batches like this, we can identify the cause and duration of all stoppages on the turbine.

In a previous study (submitted for publication), the authors performed a detailed analysis of these alarm batches and demonstrated how they can be used to reduce the strain of alarm analysis on turbine operators. Also detailed is the methodology for building a picture of which alarms are associated with which stop category when documentation is not available from the OEM.

The total duration of stoppages for each category of stop in the data is shown in figure 1. As can be seen, the stoppages related to pitch faults caused the most amount of downtime, with grid-related stoppages coming a close second. Because pitch stoppages are the most common type of fault, we focus on trying to predict these faults in this study.

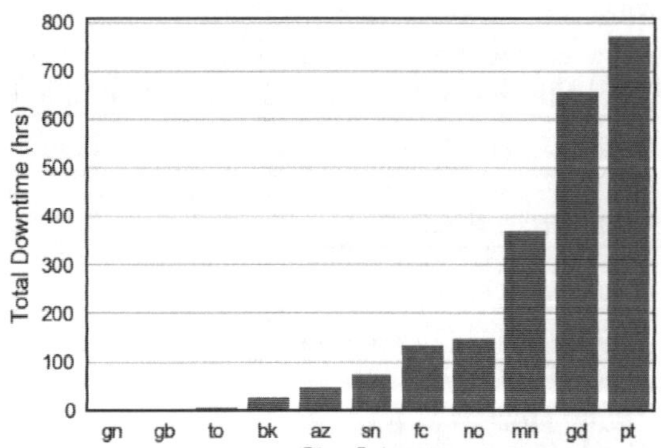

Figure 1. Total duration of stoppages for each category of stop

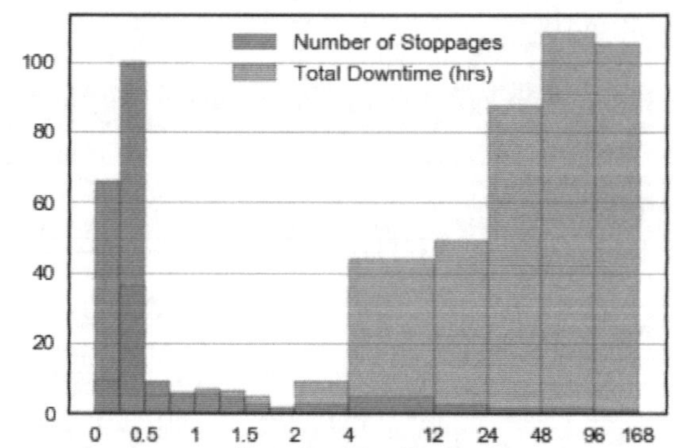

Figure 2. Distribution of pitch stoppage durations (note log scale on x axis)

A histogram showing the distribution of pitch fault durations can be seen in figure 2. It is clear that although the vast majority of stoppages are short term in nature, the longer stoppages make up the bulk of the downtime.

3.2. Suitability for Labelling Data

It was found that 69% of batches with a duration greater than 4 hours and stop category relating to a sub-system fault were represented in the maintenance logs as unplanned repairs, and in all these cases were correctly attributed to the correct sub-system. This was further confirmed by the availability data, where there are counters for planned and unplanned maintenance (though the root cause/sub-system is not given in the availability data, and there is no distinction between grid-related faults or turbine sub-system faults, etc.). All routine maintenance activities that appeared in the maintenance logs were also represented as such in the batch data.

Of the remaining batches that represented stoppages with > 4 hour durations and with a stop category relating to sub-system faults which were *not* represented in the maintenance logs, 23% were immediately preceded by a shorter-duration batch relating to either a grid fault or routine maintenance. The remaining 8% of batches were related to stoppages for a number of turbines on a single date which was due to a system software upgrade.

As well as stoppages which appeared in the maintenance logs, many shorter term (i.e. less severe) stoppages which did not appear in the logs were correctly captured. These were confirmed by cross-referencing with the availability data and SCADA power and wind speed values to confirm that the turbine was down during these periods.

Following these findings, a modification was made so that batches which followed each other in quick succession (i.e. with less than 2 hours difference between the end of one batch and the start of the next) were treated as one continuous stoppage, with the stop category being attributed to the first batch. Additionally, batches which occurred on dates that any software updates were carried out were removed from the data.

Importantly, these results mean that because all unplanned maintenance activities (i.e. "severe" faults) and stoppages due to less severe faults were correctly captured by the batch labelling process, the tedious manual step of labelling the data for fault classification can be replaced with this automated process.

4. Fault Prediction Methodology

4.1. Overview

Next, we attempt to predict the previously identified stoppages represented by the batches of alarms. The classification stage follows a typical machine learning strategy, outlined in figure 3. First, the batches are created using the alarms data. Next, these batches are overlaid on the 10-minute SCADA data, and times when the turbine was down due to the various types of faults (e.g. pitch, frequency converter, etc.), routine maintenance, grid issues, weather, or due to normal operation (cable untwisting, system tests), are identified.

Since we are only looking at pitch faults, batches with other types of faults, as well as times when the turbine was down due to grid issues, maintenance and stoppages as part of normal operation, are removed from the data for clear classification. Weather-related stoppages are kept in, as the wind and other weather-related parameters should inform when the turbine is down for these reasons, so the classifier should classify these times as fault-free. Hence, we are now left with fault-free SCADA data and pitch fault SCADA data. Batches below a certain minimum duration can be excluded from the data if this improves prediction performance. This would leave only faults that required a maintenance visit to restart the turbine, which may have a better fault signature in the data due to equipment degradation leading up to the stoppage. Fault-free SCADA samples from a certain time before the fault leading up to the point immediately preceding a fault sample are labelled as "pre-fault" data. This window of time leading up to the fault is known as the "pre-fault window" and can be varied to see if prediction performance improves.

Finally, the classifier is trained and tested in a variety of scenarios; namely, varying the minimum batch duration and pre-fault window, to evaluate performance.

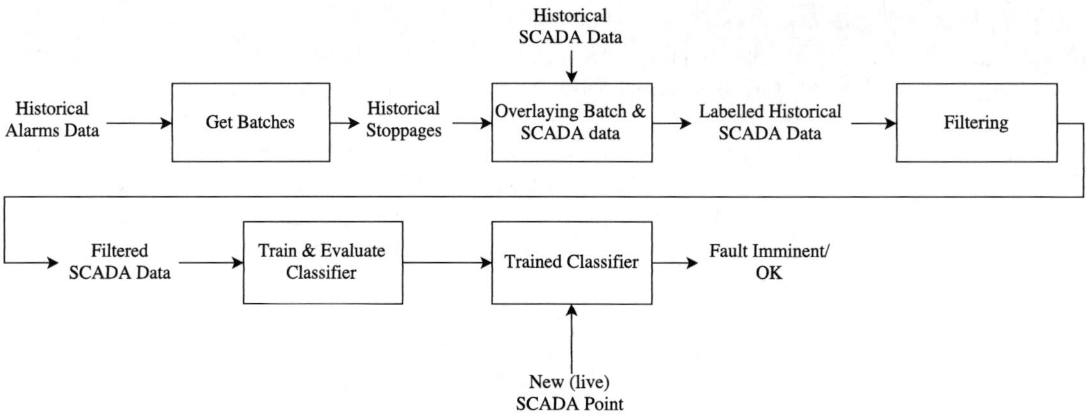

Figure 3. Overview of classification methodology

4.2. Classification Algorithm

A number of different classification algorithms were initially used on a first pass test, including support vector machines (with linear, polynomial and Gaussian kernels), decision trees and logistic regression. It was found that the support vector machines with Gaussian kernel and decision tree classifiers performed similarly, but the decision trees were much quicker to train and test and also gave human-readable rules. Ensembles of decision trees were found to improve performance, so a random forest classifier was the final type of classifier used to train all models. Details of this algorithm can be found in [20].

4.3. Feature Selection

The set of features from the SCADA data for the classification stage was based on the features selected in [14, 21, 13], with some additions. This amounted to the following features:

- Ambient temperature
- Wind speed (average, max. & standard deviation)
- Pitch system battery box temperatures
- Pitch control box temperatures
- Average set angle, actual angle and standard deviation of actual angle for each blade
- Set and actual torque
- Nacelle orientation angle

Additionally, there were the following computed features:

- The average and standard deviation of all actual blade angles
- The absolute difference between each pair of actual blade angles (e.g. blade 1 and blade 2)
- The average and standard deviation of pitch control box and battery box temperatures
- The absolute difference between each pair of control and battery box temperatures
- The absolute difference between actual and set blade angle for each blade and torque

Note that all angles were converted to radians and broken into their sine and cosine components. The absolute difference were computed from these.

4.4. Training and Testing

Models were trained for various different minimum batch durations (0, 1, 2, 4, 8, 24 and 48 hours) and fault prediction windows (2, 4, 8, 24 and 48 hours). In the training stage, the fault class was randomly undersampled such that the number of fault samples was equal to the number of fault-free samples. A cross validation strategy was used to evaluate performance whereby all but one turbine were used for training and one turbine was held out for evaluation, meaning the test set score was an average of performance across 11 different models. The testing stage of the cross validation was done in two ways; in the first, the full set of data for the evaluation turbine was used, i.e. including the class imbalance seen in real life. The second again randomly undersampled on the test set in order to benchmark against other work done in the literature. It should be noted, however, that caution should be used when directly comparing scores against previous work done as using the same methodology on different datasets can give very different scores.

5. Results

The classification performance was evaluated using both precision and recall. The formulae for calculating these can be seen below:

$$Recall = tp/(tp + fn)$$

$$Precision = tp/(tp + fp)$$

tp is the number of true positives, i.e., correctly predicted fault samples, fp is false positives, fn is false negatives, i.e., fault samples incorrectly labelled as no-fault, and tn is true negatives.

The results for all classification strategies are shown in figure 4. As can be seen, the general trends showed that the shorter the minimum duration of the pitch-fault related batches, the higher the score generally was. This implies that, because more batches were included when there was a lower minimum duration cut-off, the improved scores here are due to more fault-class data being available for training. This is consistent with findings in [13] which showed that scores were significantly improved with more data.

It should also be noted that the mean precision score was much higher when the test sets were undersampled, which is to be expected. Recall scores were not affected as much, but were still somewhat sensitive to this. On the fully sampled test set, precision scores across all minimum batch durations dramatically increased with increasing pre-fault window. This implies that precision increased with more available data. Recall seemed to be less sensitive to the window size, though in most cases in the fully sampled test set, scores were higher for shorter windows.

The scores in general were not high enough to warrant this system being deployed in the field, with a maximum precision of 0.23 and maximum recall of 0.24 on the fully sampled test set. Ideally, a high precision score would be needed to justify installing such a system - the cost to install would be relatively low as there is little to no additional hardware needed, so a low recall score and high precision score means that although not all stoppages are being predicted, a low false positive rate means that there is still a net benefit to installing the system. With a maximum of 0.23, precision was not high enough to warrant field deployment.

The precision scores were similar to scores obtained in [13] on the fully sampled set. However, scores for both precision and recall on the undersampled set were much lower than those found in [12] and [13]. This could be down to a number of factors which warrant further investigation. It is possible that the different datasets used across different studies in literature have faults that are inherently easier or harder to predict due to different turbine models or operating conditions, or due to the specific faults that occurred in the periods being analysed having stronger leading fault signatures. It should be noted, however, that when the data was shuffled before training and testing in the undersampled case, scores dramatically increased, with precision and recall in

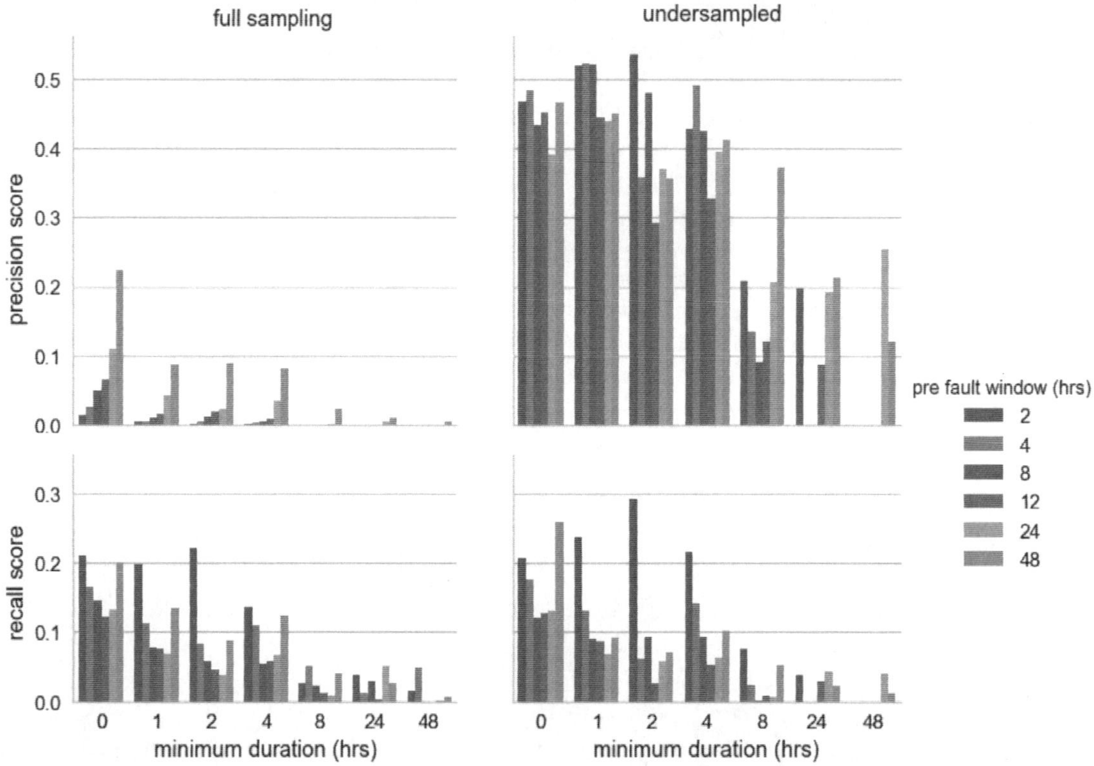

Figure 4. Overview of precision and recall scores for different sampling strategies

the high 80's and 90's. This increase is most likely due to heavy time-series effects being present in the data which artificially inflate scores. It is not clear if this step was taken in other works, which may explain the difference in scores despite similar classification approaches being used. Further investigation is needed to rule this out.

6. Conclusion

This work describes a new method to automatically identify and label every stoppage that occurred across all turbines on a wind farm using alarm system and SCADA data. This avoids the tedious step of needing to manually label stoppages in the SCADA data for fault prediction using maintenance logs. The stoppages were attributed to faults in particular subsystems, scheduled periodic maintenance events, grid-related issues, severe weather events or due to things like cable untwisting, system tests, etc. From cross referencing with maintenance logs, it was found that 92% of unplanned maintenance activities and 100% of planned maintenance activities in the logs were correctly caught. Of the unplanned maintenance activates caught, all were correctly attributed to the sub-system where the fault occurred. As well as this, many shorter-term stoppages were captured, confirmed by the turbine's availability data. Further investigations will be carried out to enable the stoppage-labelling process to be more robust to edge cases.

Fault prediction was then performed on this data using classification techniques, with various scenarios considered according to duration of the stoppages and fault prediction windows. Precision scores generally were on par with those demonstrated in literature, however recall scores were lacking. Performance dramatically increased when the data was shuffled, but these scores were artificially high due to strong time series effects.

The next steps to be taken to improve predictive performance involve creating time-lagged features to store "memory" of previous turbine states in the data. As well as this, a broader range of classification techniques will be employed to see if scores can be improved. More data will be used to see if scores can approach those seen in other works. It would be useful to benchmark all the classification techniques in literature on a unified dataset. This would also serve to rule out whether time-series effects arising from shuffling data in the cross-validation or testing stage may be inflating scores. Using the automated fault labelling process described in this research should significantly speed up the data preprocessing steps when performing classification going forward.

Acknowledgement

This research was funded by Science Foundation Ireland (SFI) Centre MaREI - Centre for Marine and Renewable Energy (12/RC/2302).

References

[1] S. Krohn, P.E. Morthorst, and S. Awerbuch, *The Economics of Wind Energy.* European Wind Energy Association, 2009.
[2] P. Tchakoua *et al.*, "Wind Turbine Condition Monitoring: State-of-the-Art Review, New Trends, and Future Challenges," *Energies*, vol. 7, no. 4, pp. 2595–2630, apr 2014.
[3] V. A. Peters, A. B. Ogilvie, and C. R. Bond, "Continuous Reliability Enhancement for Wind (CREW) Database : Wind Plant Reliability Benchmark," Tech. Rep. September, 2012.
[4] Peter Tavner, *Offshore Wind Turbines: Reliability, availability and maintenance.* Institution of Engineering and Technology, jan 2012.
[5] Eirgrid, "Industry Guide to the I - SEM," Tech. Rep., 2017.
[6] W. Yang *et al.*, "Wind turbine condition monitoring: Technical and commercial challenges," *Wind Energy*, vol. 17, no. 5, pp. 673–693, 2014.
[7] W. Yang, R. Court, and J. Jiang, "Wind turbine condition monitoring by the approach of SCADA data analysis," *Renewable Energy*, vol. 53, pp. 365–376, may 2013.
[8] J. Tautz-Weinert and S. J. Watson, "Using SCADA data for wind turbine condition monitoring a review," *IET Renewable Power Generation*, vol. 11, no. 4, pp. 382–394, mar 2017.
[9] E. Lapira *et al.*, "Wind turbine performance assessment using multi-regime modeling approach," *Renewable Energy*, vol. 45, pp. 86–95, 2012.
[10] S. Butler *et al.*, "A feasibility study into prognostics for the main bearing of a wind turbine," *2012 IEEE International Conference on Control Applications*, pp. 1092–1097, oct 2012.
[11] A. Zaher *et al.*, "Online wind turbine fault detection through automated SCADA data analysis," *Wind Energy*, vol. 12, no. 6, pp. 574–593, sep 2009.
[12] A. Kusiak and W. Li, "The prediction and diagnosis of wind turbine faults," *Renewable Energy*, vol. 36, no. 1, pp. 16–23, 2011.
[13] J. L. Godwin and P. Matthews, "Classification and Detection of Wind Turbine Pitch Faults Through SCADA Data Analysis," *International Journal of Prognostics and Health Management*, vol. 4, p. 11, 2013.
[14] A. Kusiak and A. Verma, "A data-driven approach for monitoring blade pitch faults in wind turbines," *IEEE Transactions on Sustainable Energy*, vol. 2, no. 1, pp. 87–96, 2011.
[15] K. Leahy *et al.*, "Diagnosing wind turbine faults using machine learning techniques applied to operational data," *2016 IEEE International Conference on Prognostics and Health Management (ICPHM)*, pp. 1–8, 2016.
[16] R. L. Hu *et al.*, "Using Domain Knowledge Features for Wind Turbine Diagnostics," in *2016 15th IEEE International Conference on Machine Learning and Applications (ICMLA).* IEEE, dec 2016, pp. 300–307.
[17] Y. Qiu *et al.*, "Wind turbine SCADA alarm analysis for improving reliability," *Wind Energy*, vol. 15, no. 8, pp. 951–966, nov 2012.
[18] C. Kaidis, B. Uzunoglu, and F. Amoiralis, "Wind turbine reliability estimation for different assemblies and failure severity categories," *IET Renewable Power Generation*, vol. 9, no. 8, pp. 892–899, 2015.
[19] E. Gonzalez, M. Reder, and J. J. Melero, "SCADA alarms processing for wind turbine component failure detection," *Journal of Physics: Conference Series*, vol. 753, no. 7, p. 072019, sep 2016.
[20] L. Breiman, "Random Forests," *Machine Learning*, vol. 45, no. 1, pp. 5–32, nov 2001.

[21] B. Chen *et al.*, "Wind turbine SCADA alarm pattern recognition," *IET Conference on Renewable Power Generation (RPG 2011)*, pp. 363–368, 2011.

IOP Conf. Series: Journal of Physics: Conf. Series **926** (2017) 012012 doi:10.1088/1742-6596/926/1/012012

Modelling Wind Turbine Failures based on Weather Conditions

Maik Reder and Julio J. Melero

CIRCE - Universidad de Zaragoza, C/ Mariano Esquillor 15, 50018, Zaragoza, Spain

E-mail: mreder@fcirce.es

Abstract. A large proportion of the overall costs of a wind farm is directly related to operation and maintenance (O&M) tasks. By applying predictive O&M strategies rather than corrective approaches these costs can be decreased significantly. Here, especially wind turbine (WT) failure models can help to understand the components' degradation processes and enable the operators to anticipate upcoming failures. Usually, these models are based on the age of the systems or components. However, latest research shows that the on-site weather conditions also affect the turbine failure behaviour significantly. This study presents a novel approach to model WT failures based on the environmental conditions to which they are exposed to. The results focus on general WT failures, as well as on four main components: gearbox, generator, pitch and yaw system. A penalised likelihood estimation is used in order to avoid problems due to for example highly correlated input covariates. The relative importance of the model covariates is assessed in order to analyse the effect of each weather parameter on the model output.

1. Introduction

Over the past years wind turbine (WT) operation and maintenance (O&M) has become an emerging field of research. Wind farm operators can profit substantially from shifting towards predictive maintenance strategies rather than employing corrective tasks. Understanding the wind turbine degradation in operating conditions plays an important role in this transition. Predictive maintenance models take into account age-dependent component reliability degradation. Modelling the accessibility of wind farms is usually carried out including the weather conditions at site. However, the impact of meteorological conditions on the component lifetime is mainly neglected. Yet, taking into account the environmental conditions that are more likely to provoke component failures, could enhance predictive maintenance models significantly.

Several studies have been carried out investigating the failure behaviour of wind turbines. Many different data bases have been analysed with the aim of quantifying failure rates and downtime related to the wind turbine systems and their components. Examples are given in [1–6]. The change in failure rate over time is frequently modelled using Weibull distributions, deriving the so-called bathtub curve. The latter is divided into three phases - early, random (useful life) and ageing phase. Early life failures are characterised by a decreasing rate over time, while ageing or wear-out is represented by an increasing failure rate. During the useful life period the failure rates are assumed to be constant.

IOP Conf. Series: Journal of Physics: Conf. Series **926** (2017) 012012 doi:10.1088/1742-6596/926/1/012012

Previous studies have shown that not only the turbine age, but also certain combinations of weather conditions can affect their life-time negatively. During the useful life-time, the assumption of constant failure rates does not always hold true, especially when considering shorter time-intervals, such as for example the failure occurrences on a monthly basis. There are significant variations in failure rates throughout the year that need to be taken into account by the wind farm operators. With this they can react properly to upcoming failure by initiating preventive or opportunistic maintenance actions.

Much research effort has been dedicated to identifying the critical meteorological parameters that influence the WT failure behaviour negatively. One of the first extensive analysis on the effects of weather on WT reliability was presented by Hahn et al. [7], showing increased failure rates of certain WT components with rising average daily wind speeds. In Tavner et al. [8] an annual periodicity in failure rates due to seasonal variation in weather conditions is reported, by analysing the correlation between monthly averaged wind speed conditions and component failures. The effects of wind speed on WT downtimes with regards to the turbines' energy- and time-based availability is examined by Faulstich et al. [9]. Wilson et al. [10] use artificial neural networks to show that especially the gearbox and generator are very likely to fail in changeable wind conditions. McMillan et al. [11] use Markov Chain models to calculate the seasonal component failure rates and wind farm availability based on weather conditions. In Wilkinson et al. [12] a strong relationship between high wind turbine downtime and temperature and wind speed was found. Tavner et al. [13] cross-correlate component failures with average monthly maximum and mean wind speed, maximum and minimum air temperature and average daily mean relative humidity. The studies presented by Wilson et al. in [14, 15] use Markov Chains and Monte Carlos simulations to model the WT failure behaviour with respect to certain weather conditions. Recently, Faulstich et al. [16] included wind speed in an additive Weibull failure rate model in addition to the three age-dependent phases of the bathtub curve. A study published by Slimacek et al. [17] indirectly considers the environmental conditions for modelling the rate of occurrence of wind turbine failures by taking into account the number of stops due to harsh weather conditions. They conclude that the number of stops due to harsh environmental conditions are the most important drivers of their model.

The mentioned research has proved that the failure behaviour of wind turbines and their components is strongly influenced by the meteorological conditions the systems are exposed to. Nonetheless, no models have been developed yet, to directly describe the WT failure behaviour based on combinations of external covariates. Thus, this work presents a novel approach to model the failures of WTs during the useful life including the effect of environmental conditions. The model will be applied to a 2 MW case study wind farm and the failures of the whole turbine system as well as four main components are modelled. In this context, failures are defined as events that can be associated to a component breakdown, which causes a WT stop and needs intervention such as replacement or repair.

2. Methodology

A regression model based on a generalised linear model (GLM) is applied to the data. The model is set up with a Poisson response distribution and a logarithmic link function. Subsequently, a *ridge* regression is employed to estimate the model parameters.

The *ridge* regression, see e.g. [18] and [19], is a penalised parameter estimation technique, which is frequently used to estimate the parameters of a regression problem with a high number of covariates. It introduces a penalty term on the squared ℓ_2-norm of the coefficient vector.

The penalisation helps to avoid over-fitting due to the high number of covariates. Although *ridge* regression is known to introduce a bias by shrinking the model coefficients towards zero, it is capable of significantly reducing the variance, in comparison to for example ordinary least square methods.

The Poisson regression model is given by:

$$\hat{y} = \exp(\beta_0 + \sum_{i=1}^{N} \beta_i x_i^T) \quad , \tag{1}$$

where $\hat{y}$ is the failure rate, x are the external covariates, N the number of observations and β are model parameters. Eq. 1 is fitted by maximising the penalised likelihood:

$$max_{\beta,\beta_0} \quad \frac{1}{N}\sum_{i=1}^{N}\left(y_i(x_i^T\beta + \beta_0) - e^{x_i^T\beta+\beta_0}\right) - \lambda\left(\frac{1}{2}\sum_{i=1}^{N}\beta_i^2\right) \quad , \tag{2}$$

with the number of observations N and a penalisation parameter λ. The latter dictates the degree of shrinkage. Thus, larger values for λ result in more shrinkage and higher bias. At the same time, however, with increasing λ values the variance decreases. Choosing a value of the penalisation parameter λ is realised by fitting a model with maximum regularization and subsequently decreasing it until the model becomes too complex and overfitting occurs. To ensure computationally efficient parameter estimation the *L-BFGS* solver, see e.g. [20], is used.

2.1. Case Study

The model is applied to a case study including 30 turbines located in a wind farm in Spain. The WTs are geared, three bladed and pitch-regulated machines, with a rated capacity of $2MW$ each. The terrain complexity of the site is 4, which was classified using the normative IEC 61400-12-2, [21]. The data set consists of failure logbooks, SCADA and meteorological tower (met-mast) data collected during three years of operation. At the start of the data collection the turbines were five years old. An average wind farm year is modelled, whereas the observation period is introduced into the model by means of an exposure variable called model offset. Thus, the model outcome can be considered as the rate of failure occurrence in an average operational year.

2.2. Data

In this study, the monthly failures of an average year will be modelled. In the results section the months January to December are indicated with the numbers 1 to 12. The model covariates include the monthly average wind speed (WS) and turbulence intensity (TI), and the monthly maximum wind speed (MaxWS), measured at a height of 45 meters at the wind farm met-mast. Additionally, the monthly mean relative humidity (RH), precipitation (Rain) and ambient temperature (Temp) taken from closely located weather station are included. Table 1 shows the minimum, median, mean, maximum and standard deviation values for each meteorological parameter of the input data. Additionally, as an indicator for the time operating at full capacity the monthly mean power production (PWR) taken from the turbines SCADA systems divided by the rated capacity, was chosen. In order to avoid problems due to the different covariate magnitudes, the latter are centred to a mean of 0 and standardised by dividing by their standard deviation.

In a first step the model is applied to the whole data base, without further distinguishing between the failed components. Subsequently, the failure data of four main components: the gearbox, generator, pitch and yaw system are extracted from the same set and the model is applied again. In order to analyse the importance of each input variable, the standardised model coefficients are compared. This is commonly done to interpret which of the covariates contributes the most to modelling the output and helps to see which weather conditions are important for modelling the failures.

Table 1. Statistical properties for each meteorological parameter .

Variable	Minimum	Median	Mean	Maximum	Standard Deviation
WS (m/s)	4.238	6.535	6.675	10.87	2.408
TI (-)	0.105	0.118	0.119	0.134	0.012
MaxWS (m/s)	16.33	23.29	23.13	30.89	5.250
Temp (°C)	4.414	12.18	12.59	20.92	6.020
RH (-)	0.634	0.657	0.673	0.719	0.035
Rain (mm)	5.067	37.20	39.78	94.00	30.02

2.3. Model Evaluation

The model performance is evaluated using the metrics root mean square error (RMSE), mean absolute error (MAE) and the coefficient of determination R^2. Due to confidentiality reasons regarding the failure data and in order to enable the comparison of the five data input cases, the first two metrics are displayed in a normalised form indicated by NRMSE and NMAE. The metrics are defined as:

$$RMSE = \sqrt{\frac{1}{N}\sum_{i=1}^{N}(\hat{y}_i - y_i)^2}, \qquad NRMSE = \frac{RMSE}{y_{max} - y_{min}}; \tag{3}$$

$$MAE = \frac{1}{N}\sum_{i=1}^{N}\sqrt{(\hat{y}_i - y_i)^2}, \qquad NMAE = \frac{MAE}{y_{max} - y_{min}}; \tag{4}$$

$$R^2 = 1 - \frac{\sigma(\hat{y} - y)^2}{\sigma(y)^2}; \tag{5}$$

where $\hat{y}$ is the modelled variable, y the original data value and σ denotes the standard deviation. The values y_{max} and y_{min} are the maximum and minimum of the original data.

3. Results and Discussion

A summary of the model performance metrics for the five different cases is shown in Table 2. The R^2 value ranges from 0 to 1 and indicates how shows the goodness of fit of the model to the data, where 1 indicates the best fit. It can be seen that in general the model performs well for all five failure classes.

The models for the generator and pitch system, however, showed lower R^2 and higher error values than the other ones. This indicates that for these two components additional covariates, which were not included in the model, could be of importance. This will be assessed in further studies.

Figure 1 shows the original failures (black) and the modelled data (green) for wind turbine system failures without distinguishing between their components. The values are normalised to the maximum number of failures for confidentiality reasons. As an example for a separately

Table 2. Model metrics for the different components.

Measure	All failures	Gearbox	Pitch System	Generator	Yaw System
R^2	0.7982	0.8139	0.5492	0.5623	0.8837
NMAE	0.1087	0.1034	0.2456	0.1697	0.0694
NRMSE	0.1218	0.1280	0.2894	0.2079	0.0884

modelled WT component, Figure 2 displays the original and the modelled data for yaw system failures.

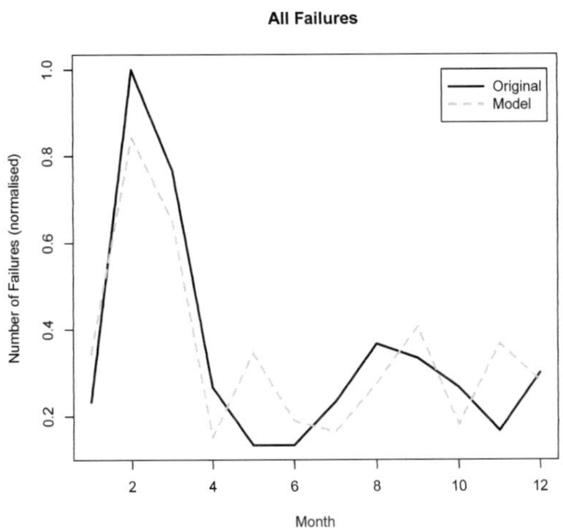

Figure 1. Model vs. data: all failures

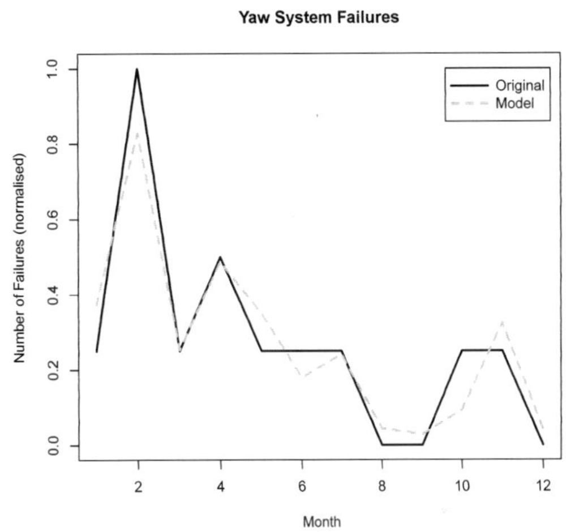

Figure 2. Model vs. data: yaw system

It can be seen that in both cases the highest failure occurrences are recorded within the second to fourth month of the year. This is consistent with previous literatures, where it is stated that failures mainly occur during the winter months, e.g. [11] and/or the transition periods between seasons, see e.g. [22].

Figures 3 to 7 show the standardised coefficient magnitudes. Positive coefficients (blue) indicate that with increasing covariates the output increases as well, while negative coefficients (orange) respectively describe the opposite behaviour. As the input data were centred and scaled, these coefficients can be interpreted as the influence of every model covariate on the dependent variable. For an easier interpretation the plots show the magnitudes as fraction of the most important regressor, which is scaled to 1.

As a multivariate model is used, the coefficients have to be interpreted with care though. The isolated effect of one regressor can theoretically only be interpreted, if the other regressors were kept constant. As this is very unlikely for environmental conditions, which in many cases are correlated, the effect should be analysed considering combinations of the meteorological parameter. Furthermore, only relative effects can be investigated.

As shown in Figure 3 the input parameters TI, $Temp$ and WS have the highest importance for modelling the data including failures of any WT component. High monthly mean turbulence intensity, high mean wind speed and low temperatures play a significant role. This is consistent

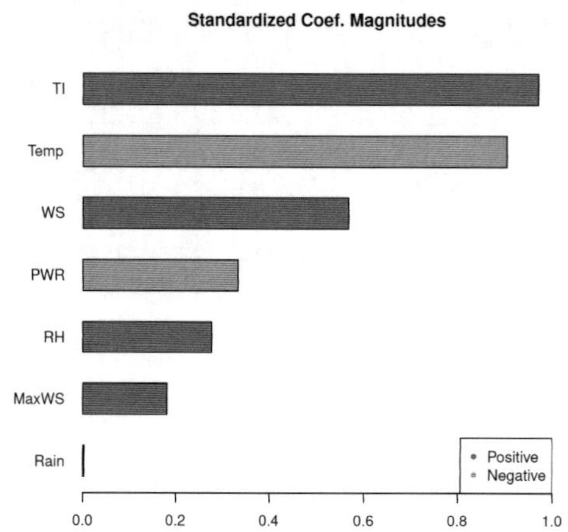

Figure 3. Variable importance: all failures

with the aforementioned literature, which stated that higher mean wind speeds and low temperatures can be correlated to higher number of WT failures. Low PWR values seem to influence the failure behaviour as well, but not very dominantly. As the active power output is usually positively correlated to wind speed, having positive coefficient magnitudes for wind speed and negative ones for the PWR variable this might seem contradictory. However, under faulty conditions the wind turbine is often performing below the expected capacity and under-performance can be seen as indicator for component failures.

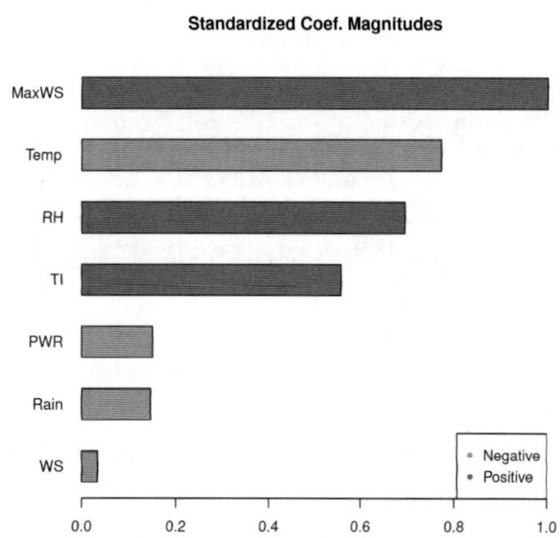

Figure 4. Variable importance: pitch system

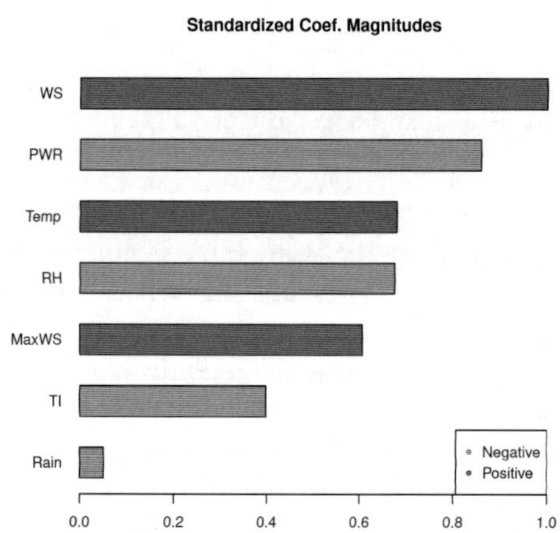

Figure 5. Variable importance: yaw system

In literature, e.g. [6] and [23], it has been shown that the different components react differently to certain combinations of environmental conditions. Thus, these should be analysed separately in order to obtain more meaningful results.

For the pitch system model (Figure 4) low temperatures and high monthly maximum wind speeds are significant. In addition to that, high relative humidity and turbulence intensity play a role when modelling the pitch system failures. These are the conditions, where the pitch system is mostly active in order to regulate the rotor speed and thus is subject to higher stresses and possible damages.

Figure 5 shows the standardised coefficient magnitudes for yaw system failures. The differences between the coefficient magnitudes are not as large as they are for the other WT components. Many meteorological factor seem to play a role in this model. A clear under-performance of the turbine can be seen, as rising wind speeds and falling *PWR* values lead to increased numbers of yaw system failures. This leads to the assumption that despite the higher mean wind speeds, the wind direction changed frequently and the yaw system had to be constantly searching for the best wind direction. Thus, a possible yaw-misalignment resulted in higher wear and under-performance. However, this should be investigated in more detail by including the wind direction in further studies.

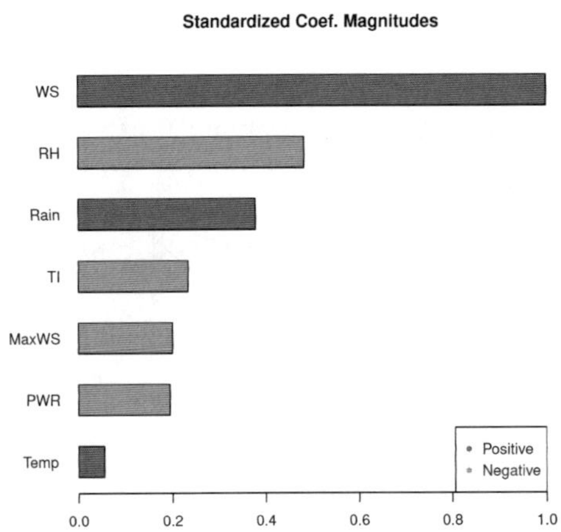

Figure 6. Variable importance: gearbox

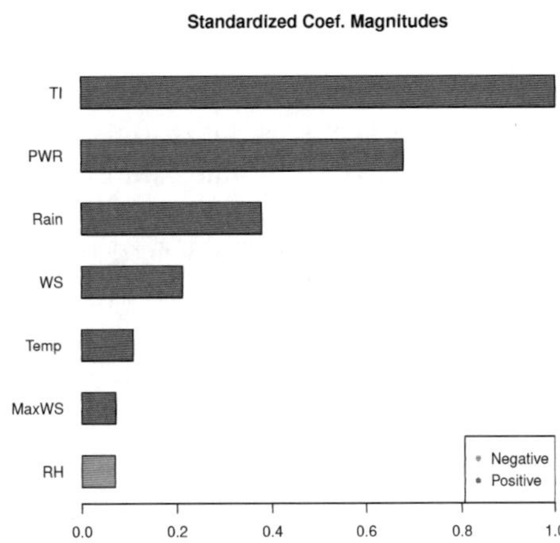

Figure 7. Variable importance: generator

Modelling the number of gearbox failures is influenced mostly by high wind speeds, as displayed in Figure 6. With higher mean wind speeds, the load on the gearbox is increasing and the component is more likely to fail.

The generator failure model (Figure 7) is mostly driven by increasing turbulence intensity and power output. Elevated *TI* introduces higher loads on the generator that has to adapt to these varying input speeds. Positive coefficients for the variables *PWR* and *WS* indicates that no under-performance was recorded before the failures. Additionally, it states that with higher power production, the generators are more likely to fail. Generator failures usually occur abruptly due to sudden changes in turbulence intensity and highly varying wind conditions, see e.g. [24]. Furthermore, the amount of precipitation plays a significant role, as water intrusion highly affects electronic equipment.

4. Conclusion

A methodology to model wind turbine failures based on the environmental conditions to which the turbines are exposed to has been introduced and applied to five failure classes. It was shown that the models work well and that it is possible to derive the relative importance of each input variable in order to analyse the effect of different covariates on the modelling output. This was used to assess which combinations of environmental conditions affected the respective component the most. Furthermore, the models have shown, that mechanical components, especially the yaw system, come along with a significant turbine under-performance. Thus, these failures could be anticipated by immediately detecting under-performance in combination with certain environmental conditions. Hence, the herein presented models serve to identify which environmental parameters influence the failure behaviour of certain wind turbine components. This information can help to anticipate failures and significantly enhance predictive maintenance models.

5. Outlook

In future studies the models will be refined and the performance of different modelling techniques and input variables will be investigated. These should be capable of handling high numbers of inputs and reduce collinearity. Further environmental variables will be tested, in order to determine the ones that mostly affect the components' failure behaviour. Instead of relative humidity, absolute humidity is expected to lead to better interpretation of the results in terms of the effect of humidity on the components. Especially yaw system failures should be modelled taking into account wind direction and wind shear.

The model will be tested on larger data sets, including more wind farms and different turbine technologies. The objective is to establish generic models describing the influence of environmental conditions on WT component failures. These can then be used for failure prediction and for assessing the environmental conditions that have the highest impact on WT components. Furthermore, these models are expected to enhance predictive O&M strategies and contribute to decreasing the overall cost of wind farms.

Acknowledgements

This project has received funding from the European Union's Horizon 2020 research and innovation programme under the Marie Sklodowska-Curie grant agreement No 642108.

References

[1] Hahn B, Durstewitz M and Rohrig K 2007 *Wind Energy* 329–332
[2] Bard J, Faulstich S and Lynding P 2011 January (Fraunhofer Inst itute for Wind Energy and Energy System Technology (IWES))
[3] Wilkinson M, Harman K, Hendriks B, Spinato F and Van Delft T 2011 *Proceedings of the European Wind Energy Conference* (Brussels, Belgium: European Wind Energy Association) pp 1 – 8
[4] Carroll J, McDonald A and McMillan D 2015 *Wind Energy* **19** 1107–1119
[5] Reder M, Gonzalez E and Melero J J 2016 *Journal of Physics: Conference Series* **753** 1–10
[6] Reder M and Melero J J 2016 *Wind Europe Conference* (Hamburg, Germany)
[7] Hahn B 1997 *FGW-Workshop Einfluss der Witterung auf Windenergieanlagen,*
[8] Tavner P, Edwards C, Brinkman A and Spinato F 2006 *Wind Engineering* **30** 55–72
[9] Faulstich S, Lyding P and Tavner P 2011 *European Wind Energy Conference* (Hamburg, Germany)
[10] Wilson G, Mcmillan D and Ault G 2012 *International Conference on Sustainable Power Generation and Supply (SUPERGEN 2012)* (Hangzhou, China)
[11] McMillan D and Wilson G 2013 *EWEA Offshore 2013* (Frankfurt, Germany)
[12] Wilkinson M, Van Delft T and Harman K 2012 *European Wind Energy Conference* April (Copenhagen, Denmark)
[13] Tavner P, Greenwood D M, Whittle M W G, Gindele R, Faulstich S and Hahn B 2013 *Wind Energy* **16** 175–187
[14] Wilson G and McMillan D 2014 *Journal of Physics: Conference Series* **524** 1–10
[15] Wilson G and McMillan D 2014 *Safety, Reliability and Risk Analysis: Beyond the Horizon - Proceedings of the European Safety and Reliability Conference, ESREL 2013* 801–809
[16] Faulstich S, Berkhout V, Mayer J and Siebenlist D 2016 *Journal of Physics: Conference Series* **749** 1–11
[17] Slimacek V and Lindqvist B H 2016 *Wind Energy* **19** 1991–2002
[18] Hoerl A E and Kennard R W 1970 *Technometrics* **12** 55–67
[19] Antoniadis A, Gijbels I and Nikolova M 2011 *Annals of the Institute of Statistical Mathematics* **63** 585–615
[20] Byrd R H, Nocedal J and Schnabel R B 1994 *Mathematical Programming* **63** 129–156

[21] IEC 2013 International Standard IEC 61400: Wind turbines Part 12-2; Ed. 1, 2013/03, Power performance of electricity-producing wind turbines based on nacelle anemometry Tech. rep. Geneva, Switzerland
[22] Guo P and Bai N 2011 *Energies* **4** 2077–2093
[23] Reder M, Yürüşen N Y and Melero J J 2018 *Reliability Engineering & System Safety* **169** 554–569
[24] T Wei 2010 *Wind Power Generation and Wind Turbine Design* (Southampton, Boston: WIT Press)

IOP Conf. Series: Journal of Physics: Conf. Series **926** (2017) 012013 doi:10.1088/1742-6596/926/1/012013

The Effect of Upscaling and Performance Degradation on Onshore Wind Turbine Lifetime Extension Decision Making

T Rubert[1], D McMillan[1], P Niewczas[2]

[1] Doctoral Training Centre in Wind and Marine Energy Systems, University of Strathclyde, 204 George Street, G1 1XW, Glasgow, UK

[2] Department of Electronic and Electrical Engineering, University of Strathclyde, 204 George Street, G1 1XW, Glasgow, UK

E-mail: `tim.rubert@strath.ac.uk`

Abstract. Ever greater rated wind turbine generators (WTGs) are reaching their end of design life in the near future. In addition, first research approaches quantified the impact of long-term performance degradation of WTGs. As a consequence, this work is aimed at discussing and analysing the impact of upscaling and performance degradation on the economics of wind turbine lifetime extension. Findings reveal that the lifetime extension levelised cost of energy ($LCOE_2$) of an 18 MW wind farm comprising of 0.5 MW rated WTGs are within the order of £23.52 per MWh. Alternatively, if the same wind farm consists of fewer 2 or 3 MW WTGs, the $LCOE_2$ reduces to £16.56 or £15.49 per MWh, respectively. Further, findings reveal that an annual performance degradation of 1.6% (0.2%) increases $LCOE_2$ by 34-41% (3.6-4.3%).

1 Introduction

Lifetime extension (LTE) is financially attractive in comparison to new investment and repowering due to the following identified root causes: the ageing of the current European wind fleet as illustrated in Figure 1; the more competitive allocation of onshore subsidies for new investments [1]; the trend of terminating repowering subsidies as observed in Germany, Spain, Denmark, and the UK [2]; the demonstration of competitive lifetime extension levelised cost of energy ($LCOE_2$) [3]; and finally the comparatively little due diligence for the lifetime extension analysis (LTEA) as well as the hard-won, long-term stakeholder relationships with local communities. As a consequence, LTE of wind turbine generators (WTGs) has become a widespread objective within todays' onshore wind energy industry.

As greater rated turbines have been commercially developed and installed in the past, there is an observable trend of greater WTG classes reaching their end of design lifetime as illustrated in Figure 2, depicting the current and future distribution of turbine classes reaching their 20th year of operation. In 2016, the distribution is predominantly composed of WTG below 0.5 MW and to a lesser impact of WTG in between 0.5 and 1 MW; however, this distribution is changing significantly over the next decade with the pre-dominant WTG class shifting towards installations in between 0.5 and 1 MW. Given this observable trend of greater scale turbines reaching their end of design lifetime, this paper is aimed at analysing the development of lifetime extension LCOE with varying turbine classes.

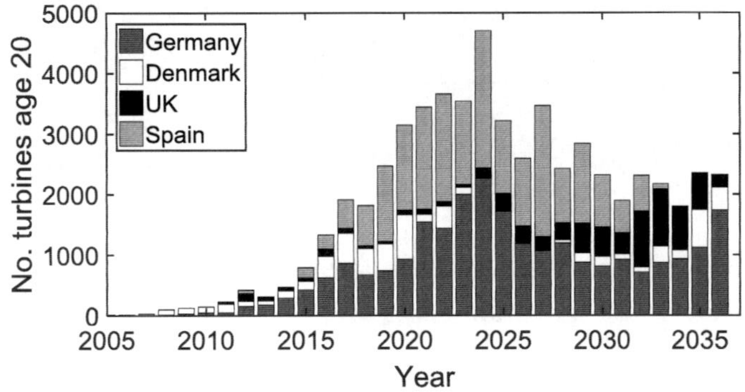

Figure 1: Number of WTGs reaching their 20th year of operation [2].

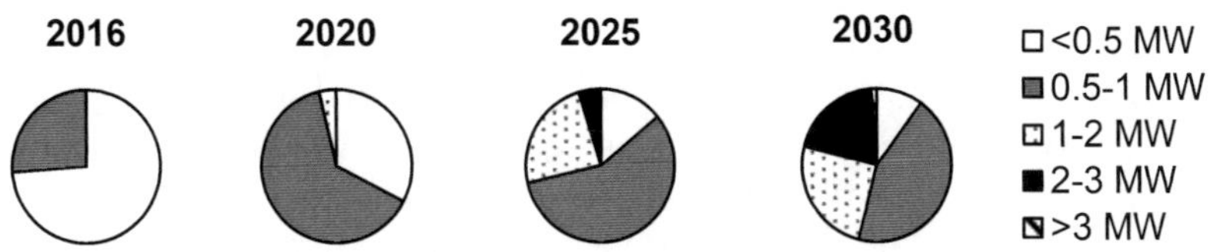

Figure 2: Distribution of WTGs reaching their end of design lifetime (20 years) [2].

Furthermore, performance degradation (PD) can be observed in all different types of machinery, with its effect mostly attributed to wear and tear of components. In 2014, Staffell and Green [4] as well as Wilkinson [5] focused on the long-term impact of PD on WTGs, with initial attempts to quantify its impact on the annual energy production (AEP) over an asset's lifetime. Given this recently published performance degradation, current LCOE models appear to omit this reduction in AEP when estimating project costs [6–9]. Consequently, the origin and effect of PD on LCOE for an asset's design life as well as for the life extended period is placed under scrutiny.

To assess the impact of upscaling and PD on lifetime extension economics, this paper's research methodology is presented in Section 2, followed by the derived model's results in Section 3. Eventually, findings are critically discussed in Section 4 before an overall conclusion is drawn in Section 5.

2 Methodology

Provided an asset has sufficient structural reserves left at the end of design life, economic LTE profitability is not guaranteed and requires thorough analysis. Therefore, based on the technical remaining useful lifetime (RUL) the aimed extension period necessitates a commercial LTEA. While Ziegler et al. [10] analysed the optimal time to switch from lifetime extension to repowering, Rubert et al. [3] developed an economic support tool to assist lifetime extension decision making. The latter is based on a two-pronged approach comprising of i) an LCOE-based decision making methodology at the end of design lifetime in conjunction with ii) a contingency-based methodology for the life-extended period (5-15 years). Based on this proposed methodology, the impact of upscaling and PD is analysed. The applied model is schematically illustrated in Figure 3, in which the authors propose to treat the lifetime extension period as a separate investment by considering the LTE expenditure as an overnight cost consisting of visual inspection, operational and loads analysis, as well as administration. Operational expenditure

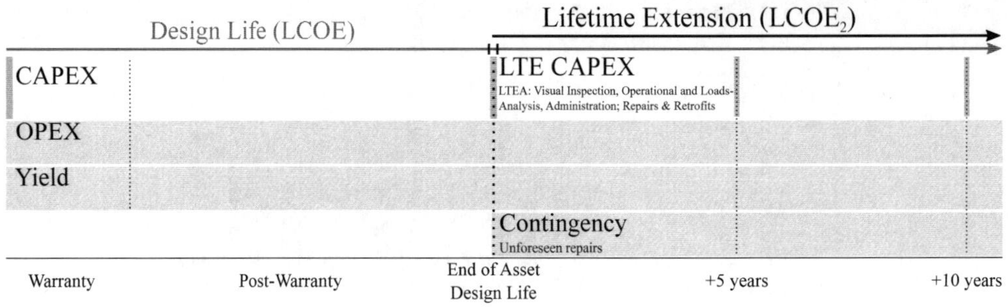

Figure 3: Economic lifetime extension model [3].

(OPEX) is modelled as a fixed annual cost, while inter-annual variability of the wind resource is not modelled. Based on the achievable cost per MWh for a given scenario ($LCOE_2$), an informed decision can be made on extending the life of an asset. Once a decision has been made to extend the lifetime, a contingency framework enables to keep a project within set economic boundaries if unforeseen repairs, retrofits or the installation of condition monitoring systems are necessary. Eventually, at the end of the life extended period; i.e., when the turbine is decommissioned, the remaining contingency fund is converted into a profit.
The model's input data is taken from central assumptions published by Rubert et al. [3]. Although, where necessary input parameters are adjusted to facilitate upscaling and PD scenarios. To allow replication of findings, adjustments to input parameters are discussed in Section 2.1 for upscaling and in Section 2.2 for PD respectively.

2.1 Upscaling

Naturally, as increasingly greater rated turbines have been commercially developed and installed, there is an inevitable fact of greater rated turbine classes reaching their end of design lifetime as illustrated in Figure 2. In order to compare the impact of upscaling of turbines on LTE economics, an 18 MW rated wind farm consisting of different generic turbine classes is modelled. The corresponding blade radii and hub heights are taken from commercially available turbines. If a significant variance in rotor size and hub height was observed at an equal rated power, the mean encountered turbine is modelled, indicated by an [M] besides individual upscaled and downscaled models, indicated by an [S]. Adjusted parameters for some turbine models are summarised in Table 1.

Table 1: Upscaling input parameters of selected 18 MW wind farm configurations. Cost estimates origin from literature [3].

Parameter	500 kW [S]	900 kW [M]	2 MW [M]	3 MW [M]
Turbines	36	20	9	6
Hub Height [m]	46.5	61.5	95.5	121.7
Blade Radius [m]	19.5	25.25	44.6	54.32
$C_{p,max}$	0.44	0.44	0.44	0.48
Wake Losses [%]	10	5.55	2.5	1.66
Wind Speed at Hub Height [m/s]	6.6	6.85	7.25	7.45
Turbulence Intensity	0.1	0.1	0.1	0.1
Weibull Shape Factor	2	2	2	2
Weibull Scale Factor [m/s]	7.44	7.73	8.18	8.41
Resulting Capacity Factor	0.25	0.27	0.37	0.39
Visual Inspection [£/Turbine]	2,150	2,150	2,688	2,688
Loads Analysis [£/Turbine]	3,500	3,500	3,500	3,500
Operational Analysis [£/Wind Farm]	59,400	33,000	15,000	10,000

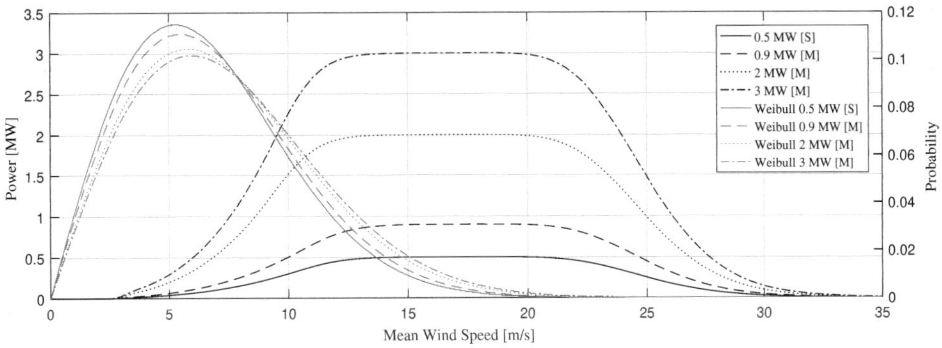

Figure 4: Simulated power curve with respect to the 10-minute mean wind speed, $P_{Sim}(v_a)$ on the left axis and wind speed probability distribution, $P_w(v_a)$ on the right axis.

2.1.1 Yield Modelling

Based on findings from [3], the site's mean wind speed is characterised by 6.85 m/s at a reference hub height of 82.5 m. Due to the natural shear profile of the wind inflow [11], each modelled turbine's mean wind speed at hub height, $U(z)$ varies. Therefore, $U(z)$ is extracted through the wind shear log-law:

$$U(z) = U(z_r)\frac{ln\frac{z}{z_0}}{ln\frac{z_r}{z_0}} \tag{1}$$

where z_r is the reference hub height, $U(z_r)$ the average wind speed at reference hub height, z the modelled turbine hub height (Table 1 and Figure 5a), and z_0 the surface roughness length (assumed as 0.03 m for open farmland, few trees and buildings) [12]. The application of the wind shear log-law assumes neutral atmospheric stability and is designed up to heights of 100 m, with evidence of inaccuracies above [13] and proposed methods to overcome this limitation [14]. The latter method requires empirical knowledge of a site, presenting challenges to adequately address in this paper, thus for simplicity the shear log-law is applied above 100 m. The annual energy production, AEP, of a wind farm is estimated by:

$$AEP = Z(1 - \eta_W)h\eta_A \int_0^\infty P_W(v_a)P_{Sim}(v_a)dv_a \tag{2}$$

where Z is the number of turbines, η_W the factor for wake induced losses, h the number of hours in a year (8760), η_A the machine availability (95% in agreement with [15]), $P_w(v_a)$ the Weibull distribution as a function of the 10-minute mean wind speed, v_a, and $P_{Sim}(v_a)$ the simulated power curve as a function of v_a [3]. The detailed yield modelling methodology of $P_{Sim}(v_a)$ can be accessed in the literature [3].

Wake losses depend on multiple factors, such as terrain topology, wind distribution, atmospheric stability, turbine thrust coefficient, C_t, spacing and the array layout [11,13,16–18]; however, in this paper the wind farm with the most turbines (36 x 500 kW) is assumed to experience a medium wake loss of 10%. Each of the modelled wind farm wake losses are scaled linearly with the number of turbines deployed in the respective wind farm.

Further, all modelled turbine topologies have the same cut-in/out wind speed while the drive train and rotor efficiency is modelled depending on the encountered efficiencies of turbines deployed in between 1997 and 2000. Figure 4 illustrates the simulated power curve with respect to the 10-minute mean wind speed (underlying standard distribution) as well as the wind speed probability distribution on the right y-axis for the selected turbine generator classes presented in Table 1.

IOP Conf. Series: Journal of Physics: Conf. Series **926** (2017) 012013 doi:10.1088/1742-6596/926/1/012013

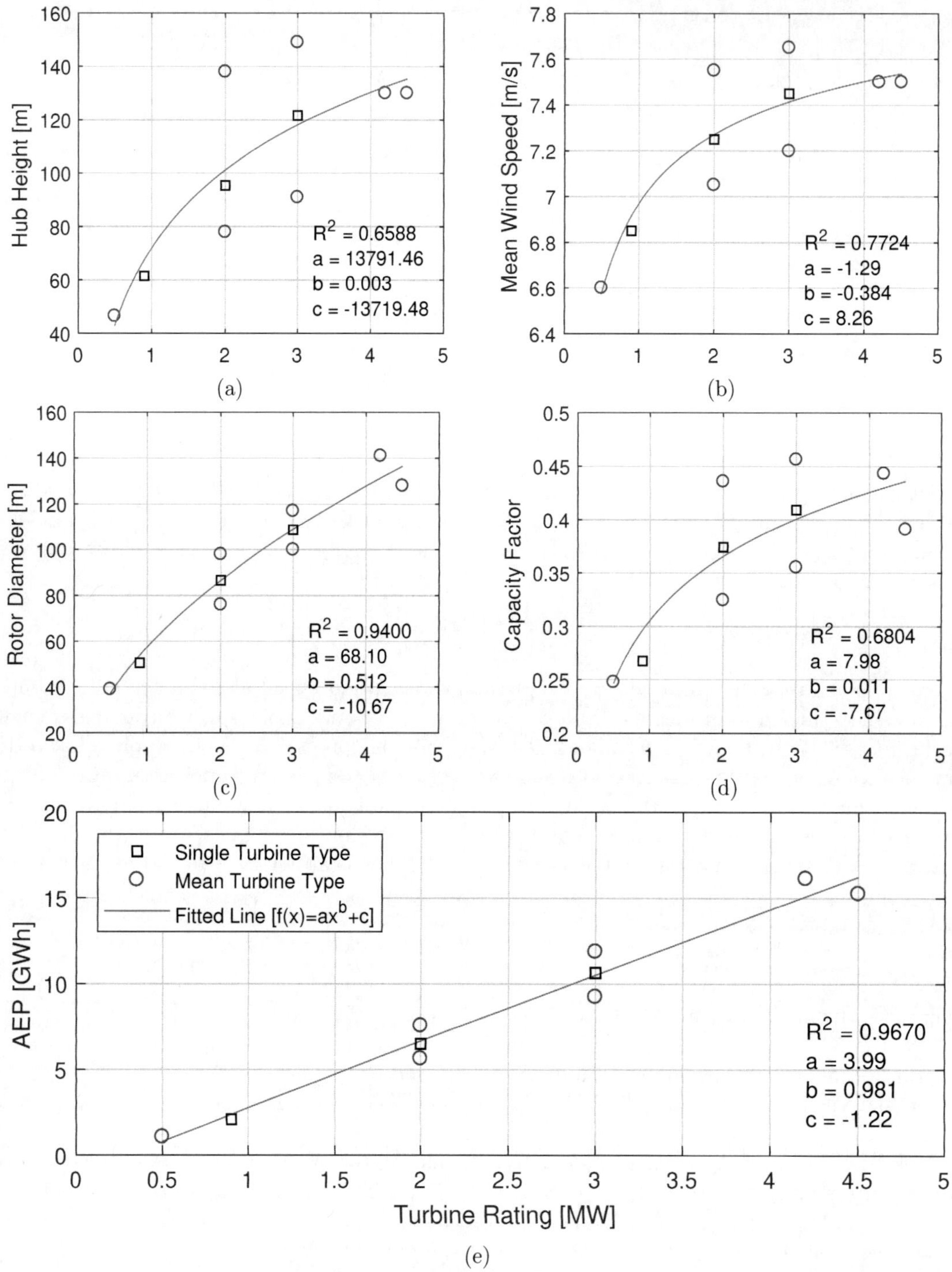

Figure 5: Modelled turbine parameters and their compound effect on capacity factor and AEP. The mean wind speed (b) is derived from the wind shear power law based on the modelled hub height (a), while the annual energy production (e) and hence capacity factor (d) is the result of the mean wind speed (b), rotor diameter (c) as well as the Weibull distribution and modelled power curve (Figure 4). A square displays an average turbine parameter, whereas a dot indicates encountered min/max parameters.

Under the presented modelling parameters, the 0.5 [S], 0.9 [M], 2 [M], and 3 MW [M] turbines achieve a capacity factor of 0.25, 0.27, 0.37, and 0.39 respectively. In addition, Figure 5 illustrates the modelled input parameters of all considered turbine types; i.e., hub height (a), mean wind speed (b), and rotor diameter (c) as well as output parameters, i.e., capacity factor (d) and AEP (e) respectively. The turbine model's annual energy production (AEP) was further compared to [19], giving confidence in the model.

2.1.2 Expenditure Modelling

It is important to highlight that for the life extended period's $LCOE_2$ estimate, the cost per MWh is dependent on the AEP, OPEX, and lastly by the LTEA's captial expenditure ($CAPEX_{LTE}$) [3]. Since the initial CAPEX is not considered in the published lifetime extension $LCOE_2$ framework as illustrated in Figure 3, the derived generation cost is thus independent of an asset's initial investment cost.

With respect to the $CAPEX_{LTE}$, the cost for visual inspection for a multi-MW turbine is assumed 25% more expensive, since the total inspection area is larger. The expenditure for the loads and operational analysis is assumed to scale linearly with the number of turbines; however, economies of scale and clustering individual turbines into cells certainly reduce costs. Since available cost figures of clustering activities are not available and highly site dependent, for ease of analysis, linearity is assumed. The cost for administration of consultants is included in the other LTEA's budgets, while the owner's administration expenditure is not included in the analysis. $CAPEX_{LTE}$ is thus modelled as:

$$CAPEX_{LTE} = Z(c_v + c_l) + c_o + c_a + c_{r,r} \tag{3}$$

where c_v is the visual inspection cost per WTG, c_l the loads analysis expenditure per WTG, c_o the operational analysis expenditure, c_a the administration expenditure, and $c_{r,r}$ the cost for necessary repairs and retrofits.

The annual OPEX, $OPEX_n$ is modelled as:

$$OPEX_n = R(C_F + C_I + C_U) + AEP_n\, C_V \tag{4}$$

where R is the asset's rated power, C_F is the fixed operations and maintenance (O&M) expenditure, C_I the insurance cost, C_U the connection and use of system charges, and C_V the variable O&M expenditure. Cost parameters are illustrated in Table 2 that are based on a deployment within the UK [3].

Table 2: Lifetime extension tool parameters [3]. Optimistic and pessimistic parameters are further presented for the supplement database [20].

Parameter	Central	Range	Unit
CAPEX			
Pre-development	100	30-240	£/kW
Construction costs	1,500	1,100-1,800	£/kW
O&M			
Fixed	30,192	22,644-37,740	£/MW/y
Variable	5.1	3.83-6.38	£/MWh
Insurance	2,226	1,669-2,782	£/MW/y
Connection/system charges	3,810	2,857-4,762	£/MW/y
Other parameters			
Discount rate	10	7.5-12.5	%
$CAPEX_{LTE}$			
Visual inspection	Table 1	±25%	£/WTG
Loads analysis	3,500	2,625-4,375	£/WTG
Operations analysis	Table 1	±25%	£/Wind Farm

2.1.3 Other Assumptions

In this work inflation in labour expenditure is not considered, whereas decommissioning costs are assumed to be equalised to the turbine's scrap value. The discount rate is selected at 10%.

2.2 Performance Degradation

"Ageing is a fact of life. Its effects are inevitable for all kinds of machinery, reducing the efficiency, output and availability of steam and gas turbines, solar PV modules, batteries and automobiles alike" [4]. Essentially, if ageing effects are observable on a wind turbine, its effects will naturally impact the overall efficiency, thus reducing the AEP and capacity factor while increasing LCOE as published in the literature [1].

In fact, wind turbine performance degradation has different distinct origins when the actual long-term wind resource is not considered or alternatively normalised for. On the one hand, there is wear and tear as well as operational errors (e.g. pitch and yaw misalignment) where the degradation process is characterised by a low rate of change. On the other hand, there is turbine downtime and hence reduced availability caused by broken parts characterised by a high rate of change (failure/sensor activation). The former can have numerous triggers such as leading edge erosion (LEE), fouling, or deformations reducing aerodynamic efficiency [21, 22], reduced system performance in components along the drive train (low and high speed shaft, gearbox, generator due to inadequate lubrification, bearing failures, or gear teeth detachment), power electronics [23] and auxiliary system [24].

When dealing with wind turbine performance degradation, there is no defined standard, i.e. what sources of PD are included and what are excluded from consideration. As mentioned previously, studies have aimed to quantify long-term performance degradation with different methodologies such as based on the UK's renewables obligation certificates (ROC) register [4] or a supervisory control and data acquisition (SCADA) system respectively [5]. Naturally, both approaches have advantages and disadvantages while considering a different combination of PD sources in the analysis, as contrasted in Table 3.

Namely, Staffell and Green's method is essentially a holistic method that is more complicated in its execution and is impacted by the potential inclusion of a bias due to increasing curtailed periods. The latter that is expected with an increasing total wind energy penetration within a

Table 3: Comparison of available PD identification approaches

Approach	Staffell and Green	Wilkinson
Data source	ROC register	SCADA
Data access	Public	Confidential
Data resolution	Monthly	10 minute (1 second)[1]
Methodology	Wind farm simulation of yield output based on NASA data	Remodelling of power curve from operational data
Complexity	High	Low
Analysis type	Holistic	Specific
Embedded PD sources	Wear and tear, operational errors (pitch and yaw misalignment), curtailment, ancillary services[2] downtime (repairs, maintenance, grid faults, etc.)	Wear and tear, operational errors (pitch and yaw misalignment)
Challenges	Correct for curtailment and ancillary services (identify)	Correct for downtime (identify)[1]
Opportunities	Introduce SCADA or Elexon data	Identify impact of curtailment, downtime, and ancillary services

[1] depends on the system and operational SCADA configuration; [2] low impact at present, expected to increase in the future [25]

network or sub-network [26].

Whereas Wilkinson's approach isolates a turbine's wear and tear as well as operational errors; the method neglects to include a turbine's downtime that impacts the long-term performance. For example, if components fail and require lengthy sourcing or the installation is restricted by weather conditions (especially offshore) a turbine will not produce power for a certain period [27] that is not necessarily extractable of SCADA data (depends on the system and operational configuration).

Consequently, it is impossible to prefer one method over the other, hence depending on the aimed application of a PD metric, one has to differentiate what sources of PD to include and which to exclude. For example, in terms of revenue and economic calculations the holistic approach appears more sensible to apply, because in this application the long-term annual yield matters that is dependent on wind farm availability, wear and tear, operational errors (yaw and pitch misalignment), as well as curtailment. Arguably, the latter factor is an operational restriction characterised by a network's local capacity limits with no direct impact on degradation.

Therefore, a sensible modification to the holistic approach would be to isolate and identify the impact of curtailment for a given wind farm or area separately. This is in essence challenging, because ROC data does not contain curtailment information; however, can be overcome if SCADA data is accessible, or alternatively through the purchase of Elexon's metered generation data [28].

Similarly, for Wilkinson's approach a downtime registry would be required to introduce the impact of repairs, maintenance, grid faults etc. Also, the recorded curtailment periods can be translated into an AEP loss over time caused by de-rating. Consequently, for both methodologies access to more detailed operational data can thus aim to identify and categorise the impact of the different identified sources of performance degradation. Knowledge on the isolated impact of different PD sources can thus help users to apply suitable PD metrics depending on the type of analysis.

With regards to available performance degradation metrics, published data varies with a recorded linear annual degradation of 0.2% as published by Wilkinson [5] with a limited dataset of the first 6 operational years while Staffell and Green's UK fleet approach observed a linear annual performance degradation of 1.6%. As highlighted by Rubert et al. [1], LCOE for the design lifetime can thus increase by up to 12.62% based on a simple model, which impact is significant and under such circumstances, essential to take into consideration.

In terms of implementing long-term performance degradation parameters into the presented economic model, two 18 MW wind farms (20 x 0.9 MW and 9 x 2 MW presented in Section 2.1) were subjected to a varying linear, annual PD. The modelled annual energy production, subjected to performance degradation is thus as follows:

$$AEP_{n,\gamma} = AEP(1 - \frac{\gamma}{100})^n \qquad (5)$$

where γ is the annual degradation factor [%] ranging from 0-2, and n is the year.

Overall, the aim was to analyse the impact of PD on LCOE and LCOE$_2$ and further evaluate if results are projectable on upscaled turbines; i.e., if the cost percentage impact of PD is comparable to greater scale WTG. Concerning the latter, as identified by Bolinger and Wiser, CAPEX undergoes cyclical variations [29] and turbines within the region of 1-3 MW have comparatively equal CAPEX/MW variations [30]. Further, sub 1 MW as well as above 3 MW substantial cost differences can be observed, although comparatively little project data was analysed. Since 900 kW is at the top of the sub 1 MW classification it is financially considered as a 1 MW turbine, hence CAPEX expenditure is comparable to a 2 MW turbine allowing a comparison of LCOE.

3 Results

3.1 Upscaling

Figure 6 presents $LCOE_2$ estimates for an 18 MW wind farm consisting of different turbine size classifications for a 5-year lifetime extension period (no repairs, reconditioning or retrofits required). In addition, data was fitted to an exponential function in the form of $f(x) = ax^b + c$ scoring an R^2 goodness of fit of 0.85 for the cost estimate and roughly 0.69 for the contingency estimate respectively. Both goodness of fit indicators illustrate the great variance within turbine classes (some turbines have comparatively upscaled rotors, presumably for low wind speed regimes), hence the exponential fit and findings require caution in their application.

Findings reveal that cost reductions are most prominent within the sub-2 MW class, with the rate of change (slope of the function) reducing above. In detail, a wind farm consisting of 36 x 0.5 MW rated turbines paired with a lifetime extension strategy of 5 years can achieve $LCOE_2$ of around £23.69/MWh. The generation cost reduces to £16.59/MWh for a 2 MW

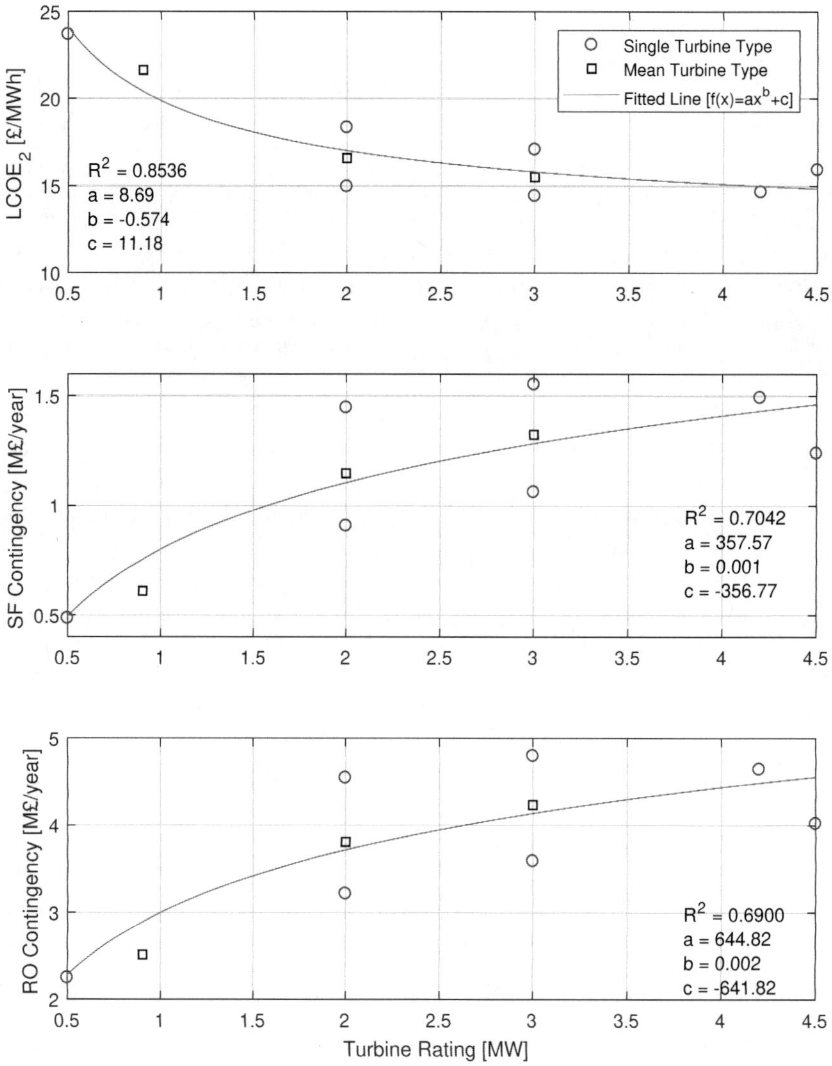

Figure 6: $LCOE_2$ +5 years upscaling results. This analysis is CAPEX independent. Subsidy-free (SF)/Renewable Obligation (RO).

turbine class, whereas a 3 MW turbine class results in $LCOE_2$ of £15.51/MWh. Within the 4 MW range there are fewer turbines commercially available; however, the Enercon E-141 with a rating of 4.2 MW (a significantly upscaled rotor diameter in this class) scores the lowest price per unit energy; i.e., £14.64/MWh.

The contingency parameter defines the maximum available budget to spend on unexpected repairs and retrofits along the lifetime extension period in order to deliver a specified operating profit [3]. In the applied model this is implemented by a defined maximum cost of energy threshold. The threshold is set in order to achieve a profit of £7.25/MWh∗, thus for the subsidy-free scenario the threshold is set £7.25 below the average day-ahead spot-market electricity price of the past 5 years [31]. For the renewable obligation (RO), the threshold is set £7.25 below the RO revenue stream defined by the 2017-2018 buy-out price and day ahead spot-market electricity price [31, 32]. The applied threshold is therefore £36.24/MWh for the subsidy-free environment and £81.82 for the RO respectively. Equally to the $LCOE_2$ results, the available annual contingency within the RO and subsidy-free framework increases with a significant rate of change below 2 MW, with an observable subsequent slowdown (Figure 6 centre and bottom graph). Besides an increase in contingency, fewer turbines are deployed, resulting in a substantial increase of available contingency per turbine.

3.2 Performance degradation

Figure 7 presents the impact of PD on the economics of lifetime extension for a wind farm consisting of 20 x 0.9 MW WTGs. The figure's x-axis represents an increasing PD corresponding to a percentage drop in AEP with the top graph presenting the impact in percent on the $LCOE_2$

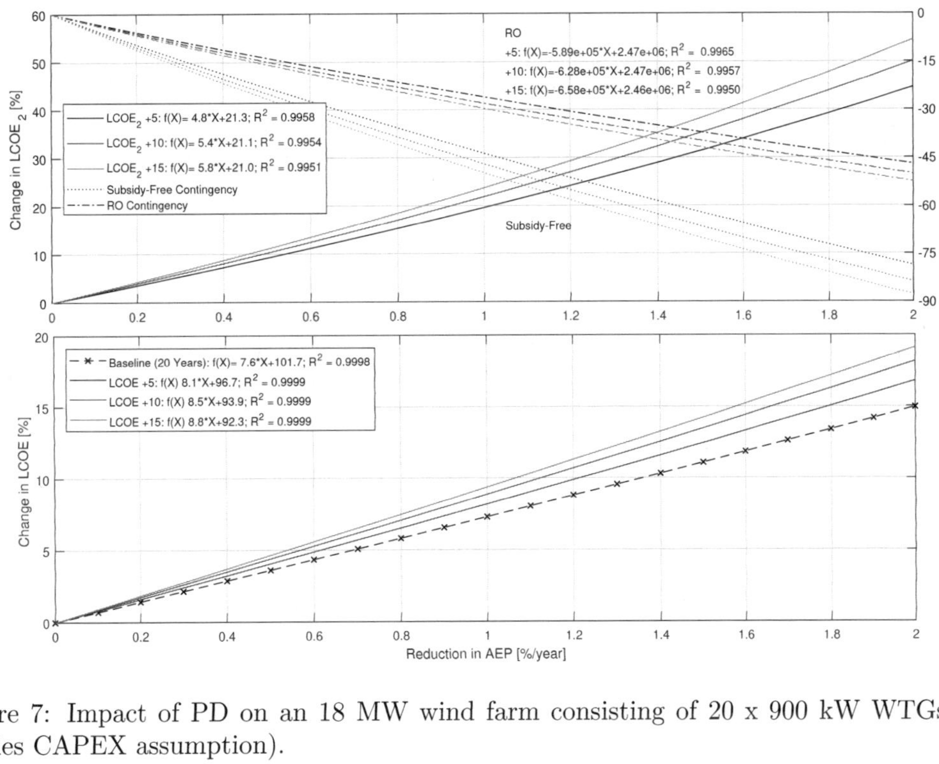

Figure 7: Impact of PD on an 18 MW wind farm consisting of 20 x 900 kW WTGs (LCOE applies CAPEX assumption).

∗This is derived from an assumed ROI of 20% based on a CAPEX of £1.6 million for a 20 year design lifetime, thus a profit of £16,000/MW/Year.

Table 4: Comparison of the impact of PD on LCOE and $LCOE_2$ of a 0.9 MW versus a 2 MW WTG in %.

Turbine Rating [MW]	Annual PD [%]	LCOE Baseline	LCOE +5	LCOE +10	LCOE +15	$LCOE_2$ +5	$LCOE_2$ +10	$LCOE_2$ +15
0.9	0.2	1	2	2	2	4	4	4
2	0.2	1	2	2	2	3	4	4
0.9	1.6	12	13	14	15	34	38	41
2	1.6	12	13	14	15	31	35	38

Table 5: Comparison of the impact of PD on annual contingency of a 0.9 MW versus a 2 MW WTG in %.

Turbine Rating [MW]	Annual PD [%]	SF +5	SF +10	SF +15	RO +5	RO +10	RO +15
0.9	0.2	-10	-10	-11	-6	-6	-7
2	0.2	-7	-8	-8	-5	-6	-6
0.9	1.6	-66	-70	-74	-39	-42	-44
2	1.6	-49	-52	-55	-36	-39	-41

metric on the left y-axis (solid line) for different life extension scenarios. A lifetime extension of 5 years is marked in black, of 10 years in blue, and of 15 years in red. In addition, the top graph presents the impact in percent on the available annual contingency under the RO (dashed line) and subsidy-free environment (dotted line) on the right y-axis. The bottom graph presents the impact of PD on LCOE for the design life of an asset (dashed black line with marker) and design life plus lifetime extension (solid line).

Concerning the latter economic metric, as argued by Rubert et al. [3], the inclusion of design life on a lifetime extension cost metric is advised against (severe discounting, asset is written off, and CAPEX dependency). Nevertheless, for the purpose of integrity results are presented. Overall, findings are further fitted to a linear function to interpolate results.

First of all, based on the applied methodology a reduction in AEP reduces yield and variable OPEX. Therefore, both effects counter-balance each other to a certain degree. However, there is a greater impact on yield, increasing the cost of energy and therefore reducing the annual contingency. Second, the impact of PD on the 20-year design lifetime is in agreement with findings from the simple analysis executed by Rubert et al. [1] with an increase of LCOE of 11.8% (1.4%) if an annual PD of 1.6% (0.2%) is encountered. Third, a comparison of the impact of PD reveals a much greater sensitivity of the $LCOE_2$ methodology (top graph) than the non-advised LCOE approach (bottom graph). In detail, results indicate that an annual PD of 1.6% (0.2%) elevates $LCOE_2$ by 34-41% (3.6-4.3%) contrary to the LCOE metric with an increase of around 13-15% (1.6-1.8%).

With regards to the impact on the available contingency of the aimed lifetime extension period, findings also reveal a significant impact. Overall, an annual PD of 1% reduces the available RO contingency by 28%, whereas in the subsidy-free environment a reduction of 47% is observed.

In addition to Figure 7, Figure 8 of the Appendix presents the impact of PD on an 18 MW wind farm consisting of 9 x 2 MW rated turbines. Findings are further contrasted in terms of the percentage impact on i) cost of energy in Table 4 and ii) contingency in Table 5. Results reveal a comparable impact with regards to LCOE and $LCOE_2$ modelling, whereas minor impact differences are observable for the RO contingency estimates. Substantial differences are observable for the subsidy-free contingency. This is due to the fact that the baseline contingency for the 900 kW turbine is substantially lower than the 2 MW's, relatively to both RO cases.

4 Discussion and Future Work

This paper's results are aimed at an asset's design life of 20 years, albeit a great share of WTG installations are nowadays designed for 25 or even 30 years of operation [33]. In such instances, upscaling $LCOE_2$ results are still valid since the LTE period is considered as a separate investment case at the end of design lifetime. Further application is however limited as modelling PD, or any conventional LCOE calculation is dependent on the design life, hence the suggested model requires modification. Therefore, LCOE and PD results are only valid for a 20 year design lifetime while $LCOE_2$ estimates are generically applicable.

Overall, the proposed upscaling and PD methodology works well if no component replacement is required. However, in deviating scenarios with component reconditioning or retrofitting, it is challenging to model turbine spare part expenditure (part of the LTEA). This is because an ex works CAPEX distribution as well as CAPEX per installed MW may not develop homogeneously with upscaling (there is evidence that the CAPEX/MW has equal price fluctuations in between 1-3 MW turbines as discussed in Section 2.2). Therefore, due diligence is required in CAPEX modelling. In addition, component replacement installation expenditure requires scrutiny as well, since greater rated cranes are neccessary with substantially higher mobilisation and daily rates. This can be further explored in order to model asset specific requirements. Nevertheless, it is expected that operators and owners will approach LTE from a more strategic point of view (+10-15 years vs. +5 years), especially for greater rated WTGs (2–3 MW) with greater budgets to spend for the LTEA (including component replacements). Given the difficulty in modelling component replacements, the available contingency may be applied to compare the necessary expenditure to the set budget.

Further, modelling PD in combination with component replacement, draws challenges to predict yield improvements (e.g., replacement of eroded blades), besides the ability to identify root causes of performance degradation down to a component level in the first place. To take this further, a turbine drive train has many components, hence sources of PD can vary greatly as highlighted in Section 2.2. Second the impact of PD may or may not be significant nor quantifiable from turbine data (reduction of 0.2% annually [5]); i.e., an annual degradation of less than 1% appears impossible to account to a specific component's output. Nevertheless research activities are executed such as for leading edge erosion by Offshore Renewable Energy Catapult (OREC) whose results suggest that the AEP can be increased by 1.5-2% following a moderate blade erosion repair [34].

With respect to the presented upscaling results, overall the methodology is applied irrespectively of differences in turbine design, hence the power coefficient with respect to the wind speed, $C_p(v)$ might deviate. In addition, a site's environmental conditions can vary significantly impacting the LCOE assessment besides modelling inaccuracies; e.g., due to the application of the log-law above 100 m height. While is it impractical to present the depth of possible modelling combinations, attached to this paper is a database allowing users to identify i) baseline LCOE, ii) baseline + lifetime extension LCOE, as well as iii) the advised $LCOE_2$ estimate for different lifetime extension scenarios (+5-15 years) under varying input parameters. The latter includes to vary i) the turbine rating, ii) the turbine and site parameters (mean wind speed, $C_{p,max}$, turbulence intensity, and Weibull shape factor), and iii) the cost scenario (central, optimistic, and pessimistic – Table 2). The tool is accessible in [20].

The UL 4143 lifetime extension standard "Outline of Investigation for Wind Turbine Generator - Life Time Extension (LTE)" [35, 36] states that individual turbines within a wind farm can be clustered into cells for the LTEA. Naturally, clustering activities decrease the $CAPEX_{LTE}$ of the load and operational analysis as well as administration per turbine. Clustering may or may not be applicable, depending on the site's external and operational conditions, as well as the applied aero elastic simulation model. In addition, the LTEA is dependent on local standards and regulations driving the cost; e.g., in Germany aero-elastic simulations are required whereas in Denmark inpections are sufficient [2]. Therefore, modelled $CAPEX_{LTE}$ that are presented in Table 1 may deviate significantly and may not be comparable to countries outside the UK. Nevertheless, sensitivities are also modelled and adjustable in the cost scenario tool attached to this paper (Table 2 of the Appendix).

With regards to the lifetime extension period, the set contingency threshold depends on an owner's or operator's operational framework as well as aimed profit margin. Further, the threshold definition is likely dependent on economies of scale; i.e., for greater rated turbines, a lower profit margin per MWh may be acceptable, based on a relatively grater AEP [MWh/MW] (Figure 5e). As a consequence, due diligence is required to evaluate a suitable threshold for a lifetime extended site.

While this paper's model applies an annual linear PD, in reality fluctuations in the rate of change are likely to be observed. Future work, could aim to identify the impact of the different sources of PD over time (Table 3) and model a case study accordingly.

Based on general industrial feedback summarised by Ziegler et al. [2] as well as discussions with operators in the field, we are of the opinion that performance degradation is likely to impact a turbine at a later stage of life than at the beginning (warranty and performance based maintenance contracts are often in place to maintain a specified availability [1]). We further think that an average fleet degradation parameter will be somewhere in between 0.2-1.6 [%/year]. Although, we believe on average, PD lies within the lower band of the given spectrum if the impact of curtailment is identified and compensated for. This is also in agreement with recent findings from Olauson et al. [37], concluding a lifetime energy loss of 6%, which according to this paper's model corresponds to an annual PD of 0.6% over a 20-year design life. Lastly, the degree of performance degradation is highly dependent on an asset's O&M procedures, thus well maintained turbines are likely less impacted by degradation processes.

5 Conclusion

This paper aims to give an overview of the impact of upscaling and PD on the economics of lifetime extension. Results presented are derived from a model with limitations, hence its application requires careful evaluation if actual conditions are comparable to the applied model input. If input data differs significantly, it is possible to replicate the tool and adapt changes according to an asset's unique requirements as this model is limited to central case assumptions. Nevertheless, to overcome this limitation, a tool is provided allowing users to adjust selected input parameters.

In summary, this paper serves as a continuation of an economic lifetime extension decision support tool [3], in order to serve the need to understand the impact of upscaling and PD as well as a combination of both on an asset. Turbine owners and operators can take these findings into consideration when subjecting a wind farm to an economic LTEA, or replicate the tool if input data differs significantly.

6 Acknowledgments

This project has received funding from the EPSRC, project reference number EP/L016680/1. We would like to thank the reviewers for their time committed on this paper.

7 Appendix

As a supplement to this paper, a database is published allowing users to adjust any combination of i) the turbine rating, ii) turbine and site parameters (mean wind speed, $C_{p,max}$, turbulence intensity, and Weibull shape factor), and iii) the cost scenario (central, optimistic, and pessimistic) [20].

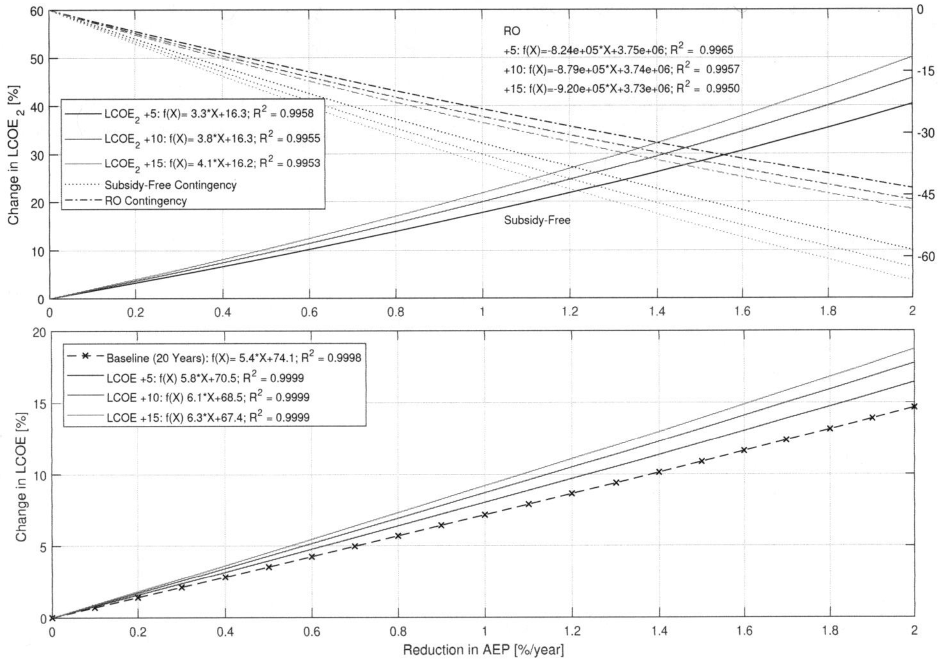

Figure 8: Impact of PD on an 18 MW wind farm consisting of 9 x 2 MW WTGs (LCOE applies CAPEX assumption).

IOP Conf. Series: Journal of Physics: Conf. Series **926** (2017) 012013 doi:10.1088/1742-6596/926/1/012013

References

[1] Rubert T, Niewczas P and McMillan D 2016 Life Extension for Wind Turbine Structures and Foundations *International Conference on Offshore Renewable Energy - CORE 2016* (Glasgow)

[2] Ziegler L, Gonzalez E, Rubert T, Smolka U and Melero J J 2018 *Renewable and Sustainable Energy Reviews* **82** 1261–1271 ISSN 13640321 URL http://linkinghub.elsevier.com/retrieve/pii/S1364032117313503

[3] Rubert T, McMillan D and Niewczas P 2017 *A Decision Support Tool to Assist with Lifetime Extension of Wind Turbines* **In review**

[4] Staffell I and Green R 2014 *Renewable Energy* **66** 775–786 ISSN 09601481 URL http://dx.doi.org/10.1016/j.renene.2013.10.041

[5] Wilkinson M 2014 Long-term Performance of Wind Farms *EWEA Workshop 2014* December pp 1–27

[6] Mone C, Hand M, Maples B and Smith A 2013 *National Renewable Energy Laboratory*

[7] Scottish Renewables 2017 An analysis of the potential outcome of a further Pot 1' CfD auction in GB A report for Scottish Renewables Tech. Rep. April URL https://www.scottishrenewables.com/publications/baringa-sr-analysis-potential-outcome-pot-1-cfd-/

[8] ARUP 2016 Review of Renewable Electricity Generation Cost and Technical Assumptions Study Report URL https://www.gov.uk/government/uploads/system/uploads/attachment_data/file/566718/Arup_Renewable_Generation_Cost_Report.pdf

[9] Wiser R, Jenni K, Seel J, Baker E and Hand M 2016 Forecasting Wind Energy Costs and Cost Drivers: The Views of the World's Leading Experts URL https://www.ieawind.org/task_26_public/PDF/062316/lbnl-1005717.pdf

[10] Ziegler L, Lange J, Smolka U and Muskulus M 2016 The decision on the time to switch from lifetime extension to repowering *WindEurope, Summit 2016* (Hamburg)

[11] Burton T, Jenkins N, Sharpe D and Bossanyi E 2011 *Wind Energy Handbook* (Chichester, UK: John Wiley and Sons, Ltd) ISBN 9781119992714 URL http://doi.wiley.com/10.1002/9781119992714

[12] Manwell J F, McGowan J G and Rogers A L 2009 *Wind Energy Explained* (Chichester, UK: John Wiley and Sons, Ltd) ISBN 9781119994367 URL http://doi.wiley.com/10.1002/9781119994367

[13] Emeis S 2014 *Meteorological Applications* **21** 803–819 ISSN 13504827 URL http://doi.wiley.com/10.1002/met.1472

[14] Gryning S E, Batchvarova E, Brümmer B, Jørgensen H and Larsen S 2007 *Boundary-Layer Meteorology* **124** 251–268 ISSN 0006-8314 URL http://link.springer.com/10.1007/s10546-007-9166-9

[15] Arwade S R, Lackner M A and Grigoriu M D 2011 *Journal of Solar Energy Engineering* **133** 041006 ISSN 01996231 URL http://solarenergyengineering.asmedigitalcollection.asme.org/article.aspx?articleid=1456211

[16] Schallenberg-Rodriguez J 2013 *Renewable and Sustainable Energy Reviews* **21** 272–287 ISSN 13640321 URL http://linkinghub.elsevier.com/retrieve/pii/S1364032112007356

[17] Johnson C, Graves A, Tindal A, Cox S, Schlez W and Neubert A 209 New developments in wake models for large wind farms *AWEA Windpower Conference* (Garrad Hassan America Inc: Chicago, IL) URL http://www.gl-garradhassan.com/assets/downloads/New_Developments_in_Wake_Models_for_Large_Wind_Farms.pdf

[18] Ali M, Matevosyan J and Milanović J 2012 *Electric Power Systems Research* **89** 70–79 ISSN 03787796 URL http://linkinghub.elsevier.com/retrieve/pii/S0378779612000399

[19] Polinder H, Van Der Pijl F F A, De Vilder G J and Tavner P J 2006 *IEEE Transactions on Energy Conversion* **21** 725–733 ISSN 08858969

[20] Rubert T 2017 Multiple Turbines Lifetime Extension LCOE Database URL http://dx.doi.org/10.15129/df362a6b-a6a9-4d60-b515-233f7286b67b

[21] Gaudern N 2014 *Journal of Physics: Conference Series* **524** 012031 ISSN 1742-6596 URL http://stacks.iop.org/1742-6596/524/i=1/a=012031?key=crossref.ce2c95aa1636e810836dfb0186948033

[22] Sareen A, Sapre C A and Selig M S 2012 *Wind Engineering* **36** 525–534 ISSN 0309-524X

[23] Le B and Andrews J 2016 *Wind Energy* **19** 571–591 ISSN 10954244 URL http://doi.wiley.com/10.1002/we.1851

[24] Wan S, Cheng L and Sheng X 2015 *Energies* **8** 6286–6301 ISSN 1996-1073 URL http://www.mdpi.com/1996-1073/8/7/6286/

[25] NationalGrid 2017 Freqeuncy Response Volumes URL http://www2.nationalgrid.com/UK/Industry-information/Electricity-transmission-operational-data/Data-explorer/Outcome-Energy-Services/

[26] Grünewald P, McKenna E and Thomson M 2015 *IET Renewable Power Generation* **9** 66–77 ISSN 1752-1416 URL http://digital-library.theiet.org/content/journals/10.1049/iet-rpg.2013.0320

[27] Jia X, Jin C, Buzza M, Wang W and Lee J 2016 *Renewable Energy* **99** 1191–1201 ISSN 09601481 URL

http://linkinghub.elsevier.com/retrieve/pii/S0960148116307194
[28] ELEXON 2017 Data Flows URL https://www.elexon.co.uk/about/other-services/data-flows/
[29] Bolinger M and Wiser R 2011 Understanding trends in wind turbine prices over the past decade Tech. rep. Lawrence Berkeley National Laboratory Berkeley
[30] Wiser R and Bolinger M 2016 2015 Wind Technologies Market Report URL https://energy.gov/sites/prod/files/2016/08/f33/2015-Wind-Technologies-Market-Report-08162016.pdf
[31] NordPool 2016 N2EX Day Ahead Auction Prices 2012-2016 Daily GBP URL http://www.nordpoolspot.com/Market-data1/N2EX/Auction-prices/UK/yearly/?view=table
[32] Ofgem 2017 Renewables Obligation (RO) buy-out price and mutualisation ceilings for 2017-18 URL https://www.ofgem.gov.uk/publications-and-updates/renewables-obligation-ro-buy-out-price-and-mutualisation-ceilings-2017-18
[33] De Vries E 2017 Enercon raises the bar URL http://www.windpowermonthly.com/article/1428420/exclusive-enercon-raises-bar
[34] OREC 2016 Catapult delivers first blade leading edge erosion measurement campaign URL https://ore.catapult.org.uk/press-release/catapult-delivers-first-blade-leading-edge-erosion-measurement-campaign/
[35] UL 2016 Outline of Investigation for Wind Turbine Generator - Life Time Extension (LTE) URL http://www.shopulstandards.com/ProductDetail.aspx?productId=UL4143_1_B_20161031(ULStandards2)
[36] Rajput V 2016 From financial death to the life Life Time Extension (LTE) *WindEurope Summit 2016* (Hamburg)
[37] Olauson J, Edström P and Rydén J 2017 *Wind Energy* ISSN 10954244 URL http://doi.wiley.com/10.1002/we.2132

IOP Conf. Series: Journal of Physics: Conf. Series **926** (2017) 012014 doi:10.1088/1742-6596/926/1/012014

The Financial Benefits of Various Catastrophic Failure Prevention Strategies in a Wind Farm: Two market studies (UK-Spain)

N Y Yürüşen[1,2]*, J Tautz-Weinert[1,2], S J Watson[2,3], J J Melero[1]
[1] CIRCE – University of Zaragoza, C/ Mariano Esquillor 15, 50018 Zaragoza, Spain
[2] CREST – Loughborough University, Holywell Park, Loughborough LE11 3TU, UK
[3] DUWIND – Delft University of Technology, Kluyverweg 1, 2629 HS Delft, Netherlands.

*Corresponding author: nyildirim@fcirce.es

Abstract. Operation of wind farms is driven by the overall aim of minimising costs while maximising energy sales. However, in certain circumstances investments are required to guarantee safe operation and survival of an asset. In this paper, we discuss the merits of various catastrophic failure prevention strategies in a Spanish wind farm. The wind farm operator was required to replace blades in two phases: temporary and final repair. We analyse the power performance of the turbine in the different states and investigate four scenarios with different timing of temporary and final repair during one year. The financial consequences of the scenarios are compared with a baseline by using a discounted cash flow analysis that considers the wholesale electricity market selling prices and interest rates. A comparison with the UK electricity market is conducted to highlight differences in the rate of return in the two countries.

1. Introduction

The costs of wind energy generation are given by the initial investment (capital expenditure – CAPEX) and the continuous costs for operations and maintenance on the wind farm (operational expenses – OPEX) [1]. Throughout the project lifetime, all Operation and Maintenance (O&M) decisions have a significant impact on the OPEX.

Part of the job of wind farm O&M engineers is to track the reliability of wind turbines and prevent catastrophic failures. Preventive replacements may result in two different outcomes under optimistic scenarios: improving the production efficiency of the wind turbines or stabilizing the functionality of the wind turbine without any change in production performance. In a case of predicted catastrophic blade failures in the investigated farm, the latter was true – costs were anticipated without any performance improvements. Current industry practice does not always consider in detail the value of timely OPEX in the overall costs of energy for wind farms. In this paper, the authors present a discounted cash flow analysis of blade OPEX in avoiding downtime and the balance between repair costs against the value of lost energy production.

The performance of a wind turbine is assessed before and after a blade replacement by comparing power output with a power curve, bearing in mind that underperformance may be affected by other possible causes such as wakes [2], icing [3], pitch [4] and yaw [5] misalignments, etc.

Potential generation revenue losses are projected in hypothetical scenarios investigating different levels of perfect and imperfect repair. Prevention of a catastrophic failure and maintenance efficiency dependent four-year cash flow analyses are performed where the value of the same maintenance decision is compared for UK and Spanish electricity markets. This comparison gives insight in the complexity of optimizing O&M if an operator manages farms in different countries in an era of regional electricity markets in Europe [6].

2. Approach

This study analyses blade replacements in a Spanish wind farm with stall regulated turbines with rated power below 1MW. Two turbines are selected: Turbine E subject to blade replacement and Turbine C as a reference without replacement. The replacement was conducted to prevent catastrophic failure and consisted of two phases: Replacement in May 2015 and final repair in September 2015. Based on service logbooks and signals from the Supervisory Control And Data Acquisition (SCADA) system, the power production performance is evaluated.

In a second step, the financial consequences of several scenarios with different timing of the repair are investigated as detailed in Table 1. UK and Spanish electricity market data analyses are performed using the half hourly ELEXON [7] imbalance price data (assuming such data to be a reasonable proxy for daily wholesale market prices[1]) and the hourly OMIE prices [9], respectively, to conduct a discounted cash flow analysis with the projected production sales. A discounted cash flow analysis is used to estimate the net present value (NPV) of the maintenance decisions projecting forward to May 2017.

The process flowchart is given in Figure 1, which illustrates the required data and filtering operations as well as financial calculations. The analysis is conducted in MATLAB and R [10].

Table 1 Investigated maintenance timings

	Scenario	Failure	Temporary repair	Final repair
Baseline	Real case	-	5/2015	9/2015
1	Delayed final repair	-	5/2015	5/2016
2	Perfect repair	-	-	5/2015
3	Repair after downtime	5/2015	6/2015	10/2015
4	Repair after excessive downtime	5/2015	-	11/2015

[1] For negative net imbalance volume (NIV) the system buy price and for positive NIV the system sell price value is taken as the market index price according to [8].

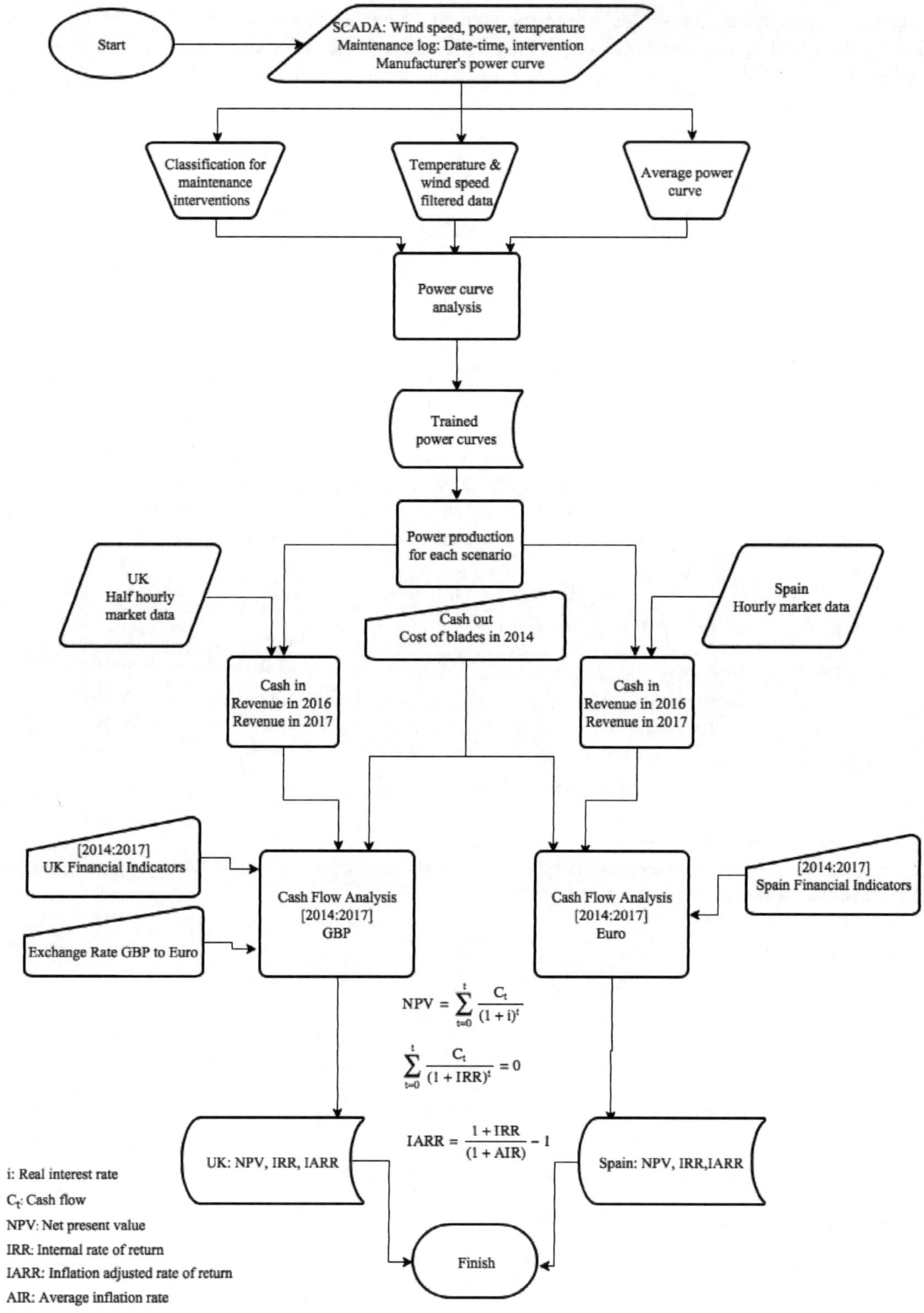

Figure 1. Flowchart of the process.

3. Power performance analysis

In an initial step, means of active power production for each wind speed bin are used to compare the two turbines. In a second step, a detailed power-curve analysis is conducted to reduce the impact of wind speed variations.

3.1. Power and wind speed bins

In Figure 2, the power production for two months (before and after the replacement) are given. The power bins of Turbine C, which was not subjected to blade replacement, tend to have a distribution with a positive skew. The power bins of Turbine E, which was subjected to blade replacement, tend to show a 'u-shaped' distribution where the turbine is spending more time at rated power. However, the reason for these differences may be found in differences in the wind speed distributions as shown in Figure 3.

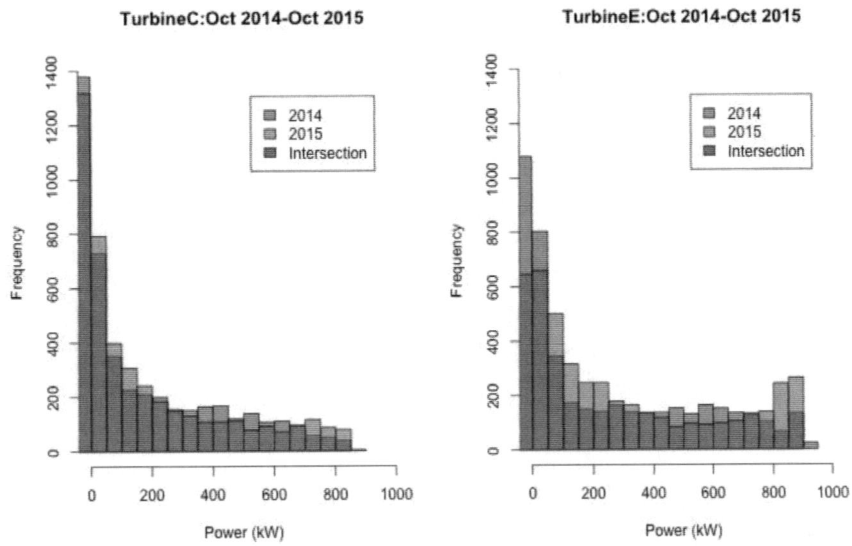

Figure 2. Power histograms for October of 2014 and 2015.

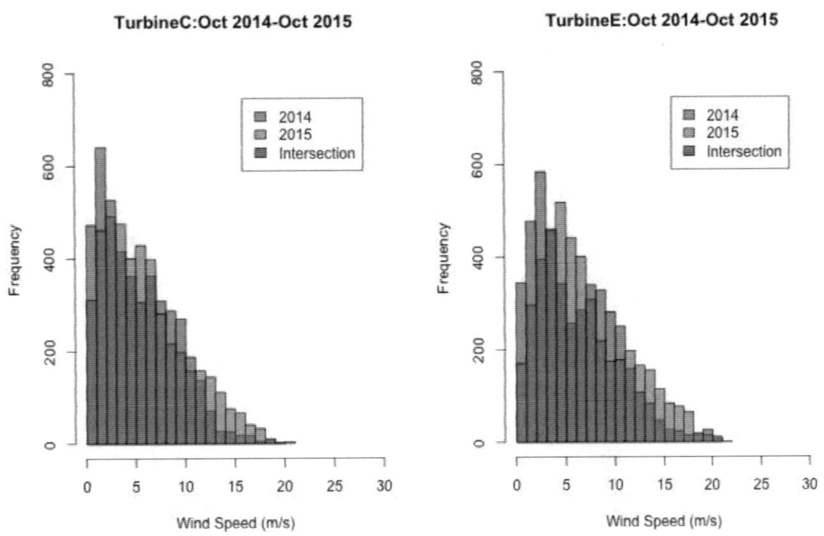

Figure 3. Wind speed distributions for October 2014 and 2015.

3.2. Power curve deviation

Although the manufacturer defines the expected performance of the turbine, the manufacturer's power curve does not account for site effects and is based on free stream wind speeds and not measurements on the nacelle. A reference power curve is derived by using the binning method described in the IEC 61400-12-1 standard [11]. In our case, 1 m/s width bins are used in order to get a reliable power curve. A representative power curve is constructed with one year of data starting after an anemometer replacement, which ensures the quality of the wind speed data. Figure 4 shows the established power curve.

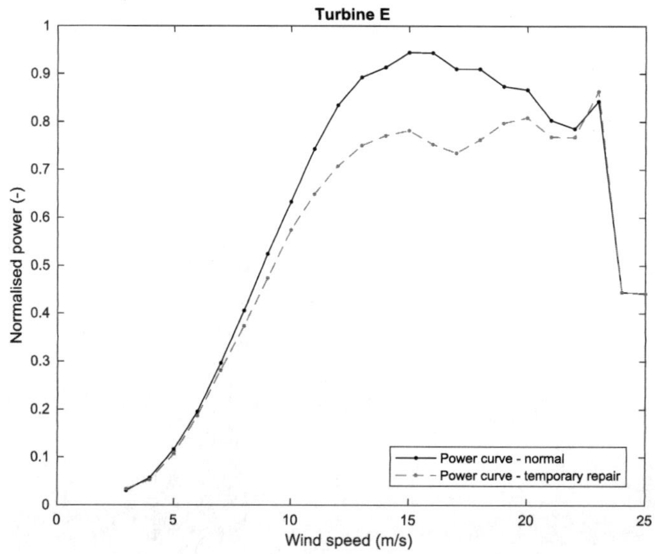

Figure 4. Wind turbine power curves based on a binning method.

In the next step, the deviation from the established power curve is analysed. Therefore, the monthly averaged power production of each wind speed bin is compared with the power curve. In Figure 5 the relative deviation is illustrated by colouring: red indicates underperformance, green implies performance according to the normal power curve and blue stands for a performance better than the power curve. If there is no significant occurrence of the particular wind speed, the bin is left blank (white). Various maintenance interventions are marked with vertical lines. It can be seen, that the performance varies significantly during the four years. The period from July 2012 to June 2013 is affected by mostly underperformance for wind speeds up to 13 m/s and performance similar to or even higher than the reference for wind speeds above 13 m/s. This might be explained by a faulty anemometer as the maintenance records indicate that the anemometer was replaced in July 2013. The period from August 2013 to July 2014 had been selected for deriving the power curve and accordingly the performance is mainly normal. The subsequent period shows similarly little deviation from the power curve except for strong underperformances in January and February 2015 at high wind speeds. It is likely that this is due to icing of the blades. Filtering of the observations to ensure an ambient temperature no less than 2°C [12], as shown in Figure 6, excludes the possible icing effects. Further underperformance is seen from May to August 2015 – the period of temporary repair. After September 2015, the power output is mostly according to the power curve except for some performance higher than expected for wind speeds of 4 to 6 m/s. This cannot be explained based on the records, but it should be noted that although the relative deviation is ≥30 %, the absolute deviation is not significant due to the operation at less than 20 % of rated power.

To further quantify the underperformance of the temporarily repaired blades, a second power curve is established for May to September 2015 and added in Figure 4 (red dashed line). It can be seen that the underperformance is strongest for wind speeds from 12 to 20 m/s. There is a slight increase of power for the 23 m/s bin, but this might be affected by the small sample size in this bin. Wind speeds of 24 or 25 m/s did not occur in this period and the power is set to the values of the normal power curve here.

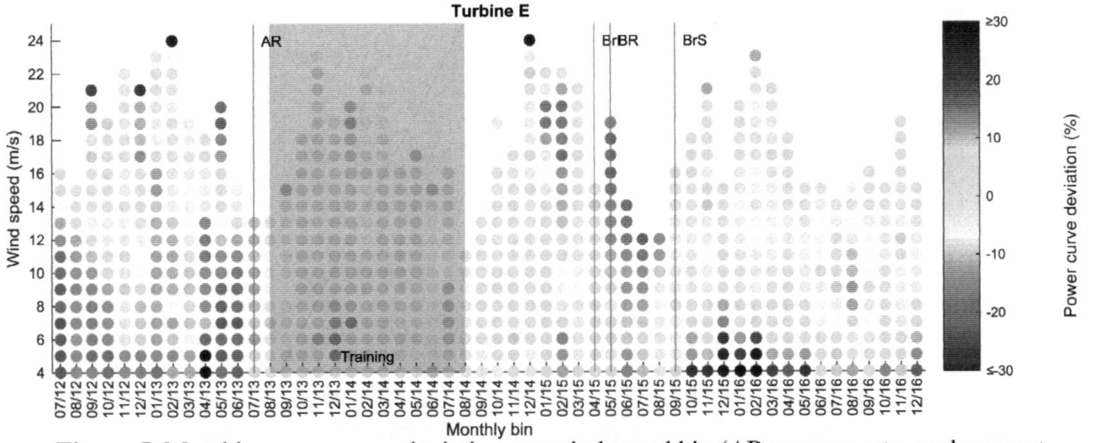

Figure 5. Monthly power curve deviation per wind speed bin (AR: anemometer replacement, Br: minor blade repair, BR: blade replacement, BrS: final blade repair on site).

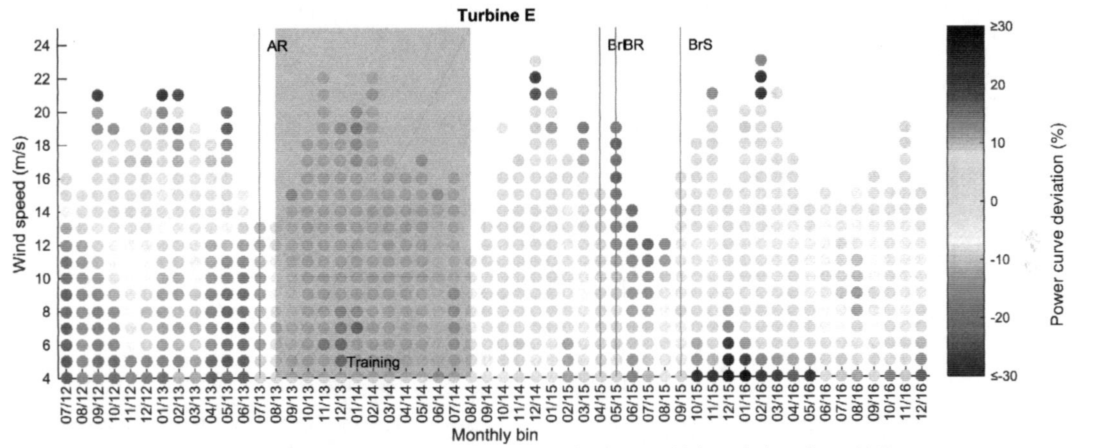

Figure 6. Monthly power curve deviation per wind speed bin – icing free (AR: anemometer replacement, Br: minor blade repair, BR: blade replacement, BrS: final blade repair on site).

4. Financial assessment

In this section, a 'what if?' analysis is performed to compare the financial implications of the different scenarios with varying maintenance dates.

4.1. Revenue evaluation

The power curves for normal operation and temporary repair are used to generate the power production in the artificial scenarios. In case of corrections to better or worse performance, only the difference of the two power curves is applied. By this, efficiency changes due to other effects and random variations remain. Table 2 shows the overall differences of the power production in the five scenarios for the first year. The power production from May 2016 to May 2017 is not varied in the scenarios (2011 MWh produced energy). In the last step the revenue is calculated by a multiplication of the market price and the energy production. The power production at 10-minute resolution is reorganised as hourly summations for the OMIE and as half hourly summations for the ELEXON data.

Table 2 Summary of power production for May 2015 to May 2016

	Scenario	Produced energy (MWh)	Change
Baseline	Real case	2360	± 0 %
1	Delayed final repair	2137	− 9.5 %
2	Perfect repair	2535	+ 7.4 %
3	Repair after downtime	2155	− 8.7 %
4	Repair after excessive downtime	1506	− 36.2 %

4.2. Cash flow analysis

Operating a wind farm is a business which requires various investments at separate times. In this manner, the profitability evaluation of possible investment is important and can be addressed by taking the project lifetime of each option into account. The discounted cash flow considers the time value of money and provides financial indicators such as net present value (NPV), internal rate of return (IRR) and inflation-adjusted rate of return (IARR). Country dynamics are here considered by using a real interest rate, which is the difference between the nominal interest rate (long term interest rate (IR)) and the inflation rate (consumer price index change (CPI) in %), as given in Table 3 [13–17]. Table 3 introduces two types of indicators as actual and forecast values. For 2017 forecasted IR and CPI values are relevant, whereas for the past years actual published IR and CPI values are appropriate. The forecasting accuracy of OECD's estimation methodology [15,17] can be tracked via 2014-2016 forecast and actual data columns.

Table 3 UK and Spain financial indicators (with NA as not available and NI as not included)

Indicator (%)	2017		2016		2015		2014	
	Forecast	Actual	Forecast	Actual	Forecast	Actual	Forecast	Actual
UK CPI	2.77	NA	0.64	0.70	0.05	0.00	1.47	1.50
UK IR	1.03	NA	1.31	1.31	1.90	1.90	2.57	2.57
UK Real IR	-1.74	NA	NI	0.61	NI	1.90	NI	1.07
Spain CPI	2.34	NA	-0.34	-0.20	-0.63	-0.50	-0.19	-0.15
Spain IR	1.71	NA	1.39	1.39	1.74	1.73	2.72	2.72
Spain Real IR	-0.63	NA	NI	1.59	NI	1.79	NI	2.87

NPV, IRR and IARR are used as evaluation indicators and calculated according to equations (1) – (3). Here, i is the minimum attractive rate of return (real interest rate in our analyses), C_n represents net cash flow at time n and N stands for the lifetime of the project. The variable i^* is the breakeven interest rate and AIR is the average inflation rate [13].

$$\mathrm{NPV}(i) = \sum_{n=0}^{N} \frac{C_n}{(1+i)^n} = \frac{C_0}{(1+i)^0} + \frac{C_1}{(1+i)^1} + \frac{C_2}{(1+i)^2} \dots + \frac{C_N}{(1+i)^N} \tag{1}$$

$$\text{If } \mathrm{NPV}(i^*) = \sum_{n=0}^{N} \frac{C_n}{(1+i^*)^n} := 0, \text{then } i^* \rightarrow \mathrm{IRR} \tag{2}$$

$$\mathrm{IARR} = \frac{1+\mathrm{IRR}}{1+\mathrm{AIR}} - 1 \tag{3}$$

Due to the complexity of real cash flow for a wind farm, only a simplified analysis is conducted here which provides a comparison of the maintenance scenarios, but does not represent the real cash flow. We consider the blade repair costs of 70,000 Euro as investment taking place in 2014 as spare blades need to be acquired before repair. In the real cash flow, the income is generated by the continuous energy sales. The income is budgeted for the initial acquisition of the turbine and all costs in operation. As we only consider one maintenance action here, we only look at a limited number of years of revenue to pay for this repair. Accordingly, we do not consider any income until May 2015. All projected energy sales from May 2015 to May 2016 of the different scenarios are utilised and registered as cash in 2016. Additionally, the real energy sales from May 2016 to May 2017 are also used to finance this repair action and registered in 2017. Any tax rates and subsidies are neglected. For the analysis of the UK setup, the initial investment is converted to GBP with an exchange rate of 1 Euro = 0.7763 GBP from May 2014 to May 2015. The cash flow analysis is subsequently conducted in GBP and finally converted back to Euro with an exchange rate of 1 GBP = 1.1828 Euro from May 2016 to May 2017 [18].

The selection of two years' energy sales as income to balance the repair costs provided a setup with at least a positive NPV as the baseline. However, the hypothesis that the full income of the two years can be used for paying the repair costs is not valid in reality. As this analysis looks only at a relative comparison, this should not matter.

Figures 7 and 8 are generated based on equation (2) and the graphical internal rate of return estimation procedure [13]. It can be seen, that scenario 2 results in an IRR of 51 % for Spain and 69 % for UK. On the contrary, scenario 4 has an IRR of 31 and 48 %, respectively for Spain and UK. An IRR ranking could be employed for the profitability evaluation of the projects [19], however, the investment evaluation must at the same time consider the NPV values for the mutually exclusive projects [20].

Table 4 gives the results of the what-if analysis showing the real NPV, which is calculated with the real interest rate, and the IARR of the four scenarios compared to the baseline. The highest NPV and IARR is found for scenario 2 in UK, the lowest for scenario 4 in Spain.

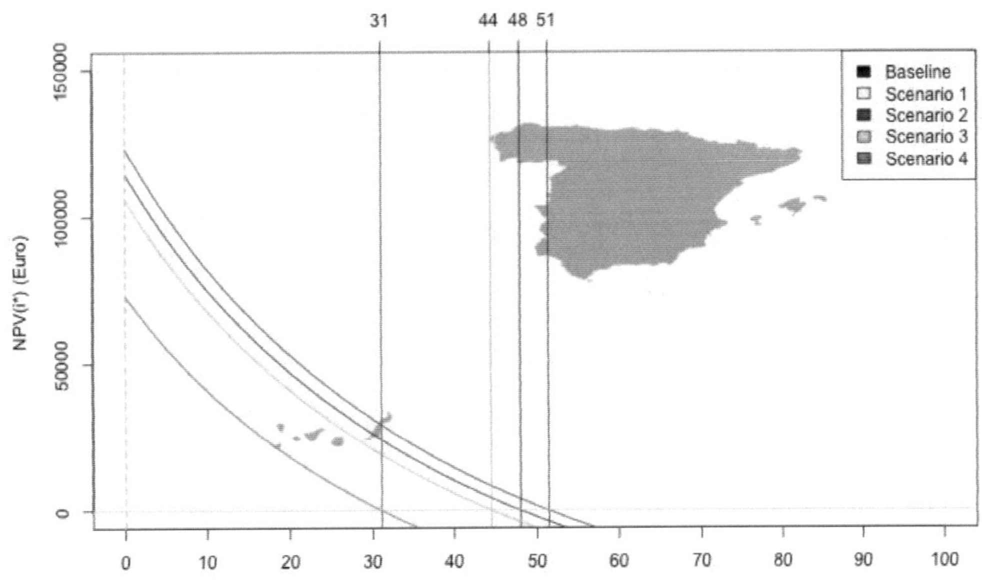

Figure 7. Graphical Internal Rate of Return Estimations for Spain (scenario 1 and 3 coincide in this figure)

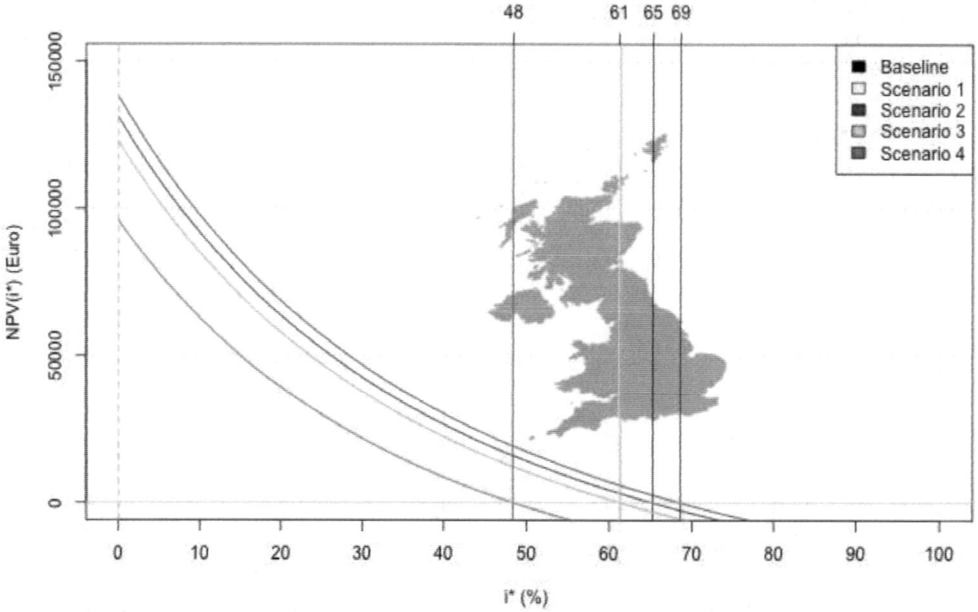

Figure 8. Graphical Internal Rate of Return Estimations for UK (scenario 1 and 3 coincide in this figure)

Table 4 Cash-flow results for the O&M scenarios (Real NPV)

	Scenario	Spain		UK		
		NPV (Euro)	IARR	NPV (GBP)	NPV (Euro)	IARR
Baseline	Real case	113,729	47.2 %	135,114	159,813	63.3 %
1	Delayed final repair	106,719	44.3 %	126,542	149,674	59.3 %
2	Perfect repair	122,021	50.6 %	142,219	168,216	66.7 %
3	Repair after downtime	105,197	43.7 %	127,017	150,235	59.6 %
4	Repair after excessive downtime	73,501	30.4 %	100,404	118,758	46.7 %

5. Discussion

The evaluation of the power production highlights that a simple comparison of energy production is not sufficient to identify underperformance due to the significant variation of the wind resource even in neighbouring turbines. Instead, a binning method has shown it is possible to generate a reasonably accurate power curve. If the deviation of the power production from the power curve is analysed, it can be seen that the performance is significantly changing during the four years. Although a filtering of events with low temperatures proves to eliminate underperformances that are related to icing, further anomalies are seen. Two main maintenance interventions, namely the anemometer replacement and the temporary blade repair, strongly influence the performance results. However, further variation is visible and might need more advanced power curve modelling to account for e.g. air density changes and directional variation. As the performance difference for the temporary repair period is in the order of up to 20 %, the limitations of the power curves will not strongly affect the evaluation of the scenarios.

The financial evaluation for the baseline case shows that a NPV of 113,729 Euro and an IARR of 47.2 % is gained under the assumption that two years of energy sales are budgeted for the repair costs (in Spain). Although this neglects the initial acquisition of the turbine and other costs, it is visible that the investment in new blades is very attractive and reasonable. The same investment in the UK would result in a higher NPV and IARR (159,813 Euro / 63.3 %), which demonstrates the differences in the electricity markets of Spain and UK. Assuming an operator with two farms in both countries and similar maintenance decisions to be made, there might be the question in prioritising one farm. Here, the importance of the blade maintenance intervention in the UK is greater from a financial viewpoint. However, this assumes no correlation between the wind conditions at the wind farms and the market prices in the two countries.

If the scenarios are compared with the baseline, we see similar trends in NPV, IARR in both countries:

1. Perfect repair is better than the baseline
2. Longer temporary repair results in higher losses
3. Excessive downtime is harmful

In detail for both countries, the perfect repair prevents five months underperformance occurrence with temporarily repaired blades and improves the NPV by 8,292 Euro (~1,658 Euro/month) in Spain and 8,403 Euro (~1,681 Euro/month) in UK. For the delayed final repair, an extension of the temporary repair period by seven months, the NPV is reduced by 7,010 Euro (~1,001 Euro/month) in Spain and 10,139 Euro (~1,448 Euro/month) in UK. The difference in the NPV change per month of temporarily repaired blades indicates that there might be potential in optimising the timing of maintenance based on monthly market and wind potential fluctuations. Further scenarios and improved power curves are needed to discuss this effect in more detail.

Additionally, we see that delayed final repair (Scenario 1) and repair after downtime (Scenario 3) cases result in nearly the same energy losses and NPV values for both countries. This could mean that a short downtime of less than a month in summer might be a preferable to longer underperformance.

6. Conclusions

A catastrophic failure prevention in a Spanish wind farm is studied in terms of the effects to the power production and financial consequences of the undertaken maintenance for various scenarios.

The power curve analysis shows that the performance of the investigated wind turbine is clearly reduced during a phase of temporary repair and regains a normal level after the final repair. Further performance variations are successfully traced back to an anemometer fault and icing events. A second power curve is established to describe the underperformance during temporary repair.

The financial consequences of the maintenance intervention are implemented in a simplified cash flow analysis assessing the rate of return of blade replacement costs balanced with revenue from power production sales of two years. The results show that the investment in new blades is financially reasonable as an inflation adjusted rate of return of 47.2% is achieved in Spain baseline case. A variation of the timing of the temporary and final repair confirms the importance of maintenance efficiency and highlights the importance of scheduling maintenance and underperformance. The comparison of the scenarios indicates that there might be potential for further optimisation of the maintenance scheduling.

A comparison of the same maintenance scenarios implemented in the UK electricity market shows that a similar investment in a UK wind farm would be favourable due to the higher electricity prices. The higher revenue is reflected in higher net present values and rate of returns. All trends of the different scenarios are the same as for the Spanish electricity market.

This study does not consider the full complexity of power performance and the cash flow for a wind farm. Future work will address the simplifications and assumptions to get a more realistic picture of decision making in the O&M of wind turbines. Additionally, a study of further fine-tuned scenarios might be beneficial to evaluate practical options and optimisation potential in more detail.

Acknowledgements

This project has received funding from the European Union's Horizon 2020 research and innovation programme under the Marie Skłodowska-Curie grant agreement No 642108 (project AWESOME -

IOP Conf. Series: Journal of Physics: Conf. Series **926** (2017) 012014 doi:10.1088/1742-6596/926/1/012014

http://awesome-h2020.eu/). The authors thank CETASA for the data. Particular gratitude is expressed to Javier Gracia Bernal, Lucas García Pérez, Marta Heras Heras and Miguel Angel Hernández Lucas for their friendly guidance. The authors thank Dr. Hasan Engin Duran for his comments on the financial indicator selection stage of the study.

References

[1] Wiser, R and Bolinger M 2017 [accessed 2017 Sep 25] *2016 Wind Technologies Market Report* US Department of Energy available from https://emp.lbl.gov/wind-technologies-market-report/

[2] Mittelmeier N, Blodau T and Kühn M 2017 Monitoring offshore wind farm power performance with SCADA data and an advanced wake model *Wind Energy Sci.* **2** pp 175-87 doi:10.5194/wes-2-175-2017

[3] Davis N N, Byrkjedal Ø, Hahmann A N, Clausen N-E and Žagar M 2016 Ice detection on wind turbines using the observed power curve *Wind Energy* **19** pp 999-1010 doi:10.1002/we.1878

[4] Bi R, Zhou C and Hepburn DM 2017 Detection and classification of faults in pitch-regulated wind turbine generators using normal behaviour models based on performance curves *Renew Energy* **105** pp 674-88 doi:10.1016/j.renene.2016.12.075

[5] Marathe N, Swift A, Hirth B, Walker R and Schroeder J 2016 Characterizing power performance and wake of a wind turbine under yaw and blade pitch *Wind Energy* **19** pp 963-78 doi:10.1002/we.1875

[6] Karova, R 2011 Regional electricity markets in Europe: Focus on the Energy Community. *Utilities Policy* **19** pp 80-6 doi:10.1016/j.jup.2010.10.001

[7] Elexon, 2017 [accessed 2017 Sep 27] *Elexon portal* available from https://www.elexonportal.co.uk

[8] Staffell,I 2016 Supplementary Material for: Measuring the Progress and Impacts of Decarbonising British Electricity *Energy Policy* **102** pp 463-75 doi: 10.1016/j.enpol.2016.12.037

[9] OMIE 2017 [accessed 2017 Sep 27] *Market results* available from http://www.omie.es/en/inicio

[10] R Core Team 2017 [cited 2017 Jan 1] *R: A language and environment for statistical computing* available from https://CRAN. R-project.org/

[11] International Electrotechnical Commission Standard 2005 *IEC 2005 Wind turbines-Part 12-1: power performance measurements of electricity producing wind turbines*

[12] Measnet 2009 [accessed 2017 Sep 29] *Power performance measurement procedure Version 5* available from: http://www.measnet.com/wp-content/uploads/2011/06/power5.pdf

[13] Park, C S 2007 *Contemporary engineering economics* (New Jersey: Pearson Education)

[14] Organisation for Economic Co-operation and Development 2016 [accessed 2017 Jan 1] *Long-term interest rates* available from https://data.oecd.org/interest/long-term-interest-rates.htm

[15] Organisation for Economic Co-operation and Development 2016 [accessed 2017 Jan 1] *Long-term interest rates forecasts* available from https://data.oecd.org/interest/long-term-interest-rates-forecast.htm

[16] Organisation for Economic Co-operation and Development 2016 [accessed 2017 Jan 1] *Inflation (consumer price index)* available from https://data.oecd.org/price/inflation-cpi.htm

[17] Organisation for Economic Co-operation and Development 2016 [accessed 2017 Jan 1] *Inflation forecast* available from https://data.oecd.org/price/inflation-forecast.htm

[18] European Central Bank 2017 [accessed 2017 Sep 29] *ECB euro reference exchange rate: Pound sterling (GBP)* available from https://www.ecb.europa.eu/stats/policy_and_exchange_rates/euro_reference_exchange_rates/html/eurofxref-graph-gbp.en.html

[19] Lang H J and Merino D N 1993 *The selection process for capital projects* (New York: John Wiley & Sons)

[20] Berk J B, DeMarzo P M and Harford J V 2008 *Fundamentals of corporate finance* (Boston: Pearson Education)

IOP Conf. Series: Journal of Physics: Conf. Series **926** (2017) 012015 doi:10.1088/1742-6596/926/1/012015

Auction-theoretic analyses of the first offshore wind energy auction in Germany

J Kreiss[1,2], K-M Ehrhart[1,2] and A-K Hanke[2]

[1]Takon GmbH, Haid-Und-Neu-Strasse 7, 76131 Karlsruhe, Germany
[2]Karlsruhe Institute of Technology (KIT), Institute of Economics (ECON), Germany

E-mail: kreiss@takon.com

Abstract. The first offshore wind energy auction in Germany led to a striking result. The average award price was 0.44 ct/kWh and even more interesting, 3 out of 4 awarded projects had a strike price of 0.0 ct/kWh. That implies that those projects will only receive the actual wholesale market price for electricity as revenue. Although there has been a strong decline in costs of offshore wind projects, such a result is still surprising. We analyzed this result auction-theoretically and showed how the auction design and the market environment can explain part of the outcome. However, another aspect of the explanation is the high risk that the awarded bidders take regarding the future development of both the project costs and the wholesale market price.

1. Introduction

Since 2017 all renewable energy support (RES) in Germany is determined in auctions [1] in accordance to the EU state aid guidelines [2]. The results of the first offshore wind energy auction in April 2017 were astonishing and attracted attention worldwide [3]. Out of the four awarded projects, three bids were 0.0 ct/kWh [4]. The biggest awarded project was the EnBW Project 'He Dreiht' with 900 MW a bid of 0.0 ct/kWh. The remainder of 590 MW was awarded to three projects of DONG Energy (now Ørsted) of which two also bid 0.0 ct/kWh and one 6.0 ct/kWh. All of those projects are located in the North Sea.

Although record low auction results for offshore wind energy were obtained in Denmark and the Netherlands in 2016, such an outcome was totally unexpected. Therefore, we analyze the reasons and implications of this result auction-theoretically. Furthermore we evaluate whether the current auction design fits the altered conditions and which consequences they have regarding the future support of renewable energy. Beside these questions we will analyze further interesting aspects of the first auction round.

2. Model

In addition to the already existing economic and political analyses of this topic, we will focus on the auction- and game-theoretic implications of the design, bidding strategies and the outcome of the first offshore wind energy auction in Germany. We will not discuss whether it is possible to economically run an offshore wind park at the given prices but whether it is reasonable to submit a 0.0 ct/kWh bid under the given conditions. We will analyze the auction regarding economic indicators such as efficiency and auction-specific risks for bidders and auctioneer like the Winner's Curse [5, 6] and non-realization [7].

For the further analyses, we introduce a consistent nomenclature of the used terms. Besides the auctioneer, the bidders $i = 1, ..., n$ participate in the auction. Each bidder represents one offshore wind project. Each bidder has specific project costs c_i which represent the individual generation costs for each unit of electricity produced from this project. The support payment is limited to $T = 300$[1] months, each month is denoted with $t = 1, ..., 300$. Furthermore, each bidder i has an expectation regarding the average electricity wholesale market price e_i. The actual wholesale market price is denoted p^t for each month t and the average over the 25 years of support is P. Moreover the projects are characterized by their nominal size s_i and their minimum bid quantity s_i^{min}. Finally, each bidder participates in the auction with a bid b_i and if a bidder is awarded, his corresponding bid determines the respective support level. The actual support payment is denoted v_i^t and is depending on the month t and awarded bidder i.

3. Analyses

The first question that arises is how the sudden drop from 5.0 ct/kWh in recent auctions for offshore wind energy in Denmark and the Netherlands to 0.0 ct/kWh in such a short time can be justified. The explanation is given by the chosen remuneration type of a sliding feed-in premium. The awarded bidders are obligated to sell generated electricity directly to the market. They receive an additional payment of the difference between their bid price (which determines their sliding premium) and the average wholesale market price as long as this difference is positive. Therefore the revenue of an awarded bidder i in month t is

$$r_i^t = \max(p^t, b_i) \tag{1}$$

and thus the respective actual support payment is

$$a_i^t = \max(0, b_i - p^t). \tag{2}$$

That is, the awarded bidder receives at least the sliding premium and can increase his revenue if the market price is above this value. This remuneration type was chosen to reduce the bidders' uncertainty regarding the future electricity market price [8]. The principle of the sliding feed-in premium is illustrated in Figure 1.

The orange line represents the wholesale electricity market price and the blue dashed line the award price. The resulting orange area corresponds to the revenue the bidder receives from selling the generated electricity to the market. Additionally the bidder obtains the support payment corresponding to the green area which he receives as the market price is below the award price and is represented in (1) by r_i^t. In contrast to the Contract for Difference (CfD) auction in the United Kingdom, revenues above the award price do not have to be paid back to the auctioneer [9]. The bidders have an asymmetric opportunity to increase their revenues in case of market prices above the award price without fearing the risk in case of low market prices, as in this case they would still receive the award price as a reliable payment.

As a result, bidding 0.0 ct/kWh means that the respective project can be run economically entirely on basis of the average wholesale market price. This implies that from an auction-theoretic point of view bids in a specific interval are dominated. This means that, independent of the behavior of the other bidders, it is never optimal for the bidder to bid in this interval.

[1] The support period is 25 years.

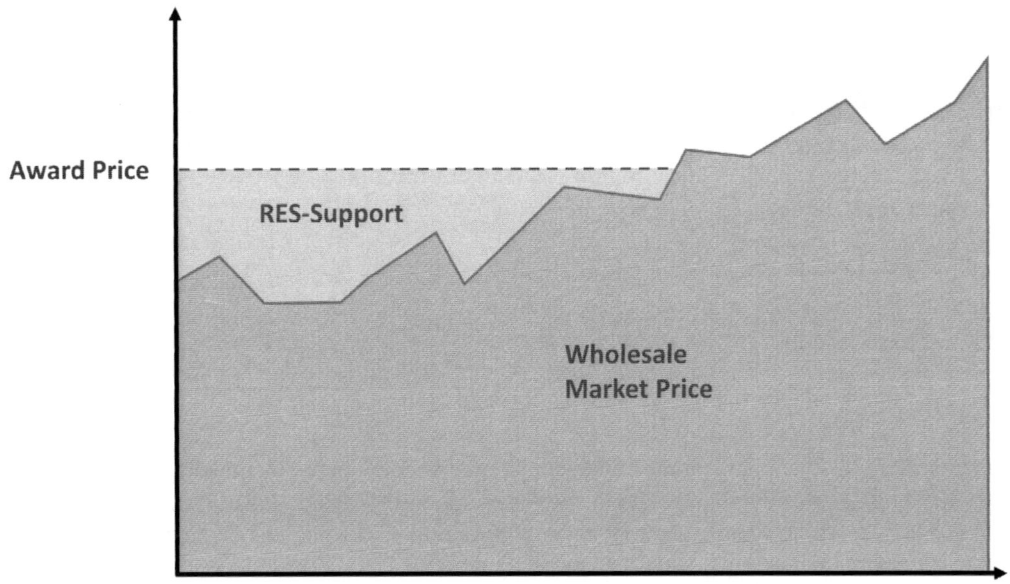

Figure 1. Illustration of the concept of a sliding feed-in premium.

If, for example, a bidder expects the average market price to be at 5.0 ct/kWh, all bids greater than 0.0 ct/kWh and lower than or equal to 5.0 ct/kWh are weakly dominated by a 0.0 ct/kWh bid. The bidder receives the same payment independent of the height of his bid in this interval as the sliding premium is lower than or equal to the market price. That is, for all bids b_i between 0.0 ct/kWh and 5.0 ct/kWh and an expected wholesale market price of $e_i = 5.0$ ct/kWh, $b_i - e_i$ is always negative and thus $r_i^t = 0.0$ ct/kWh. However, the probability of award is higher the lower the bid is. This would be the right side of Figure 1, where there is no green area.

That renewable energies are already competitive with conventional energy sources could not be foreseen when designing the auction, and therefore bidders and auctioneer face new challenges. The sliding premium was chosen to shift the risk of wrong prediction of the future market price from the bidders to the auctioneer as he faces the whole risk of unfortunate market price development. It was assumed that the award price of the auction would be higher than the actual market price for most months and thus a sliding feed-in premium gives the bidders a higher reliability than a fixed feed-in premium. However, in the light of such low bids it has the opposite effect.

The bidders face a double risk. They must predict their project costs for offshore wind parks that go online not earlier than in five years and additionally the future market price. The future market price is identical for all bidders and a high percentage of the project costs is depending on the costs for the wind turbines that are in principle available to all bidders at similar prices. Thus, auction-theoretic, both variables include a high common value component [5]. This does not imply that all bidders have the same costs to realize the projects but that those costs are interdependent. Although some bidders may get better turbine prices due to large contracts, the prices will still be similar and the electricity generation of the same turbine at the same location is the same for different companies.

In such a case, there is always the risk of the Winner's Curse which occurs when the bidder who underestimated the costs the most and did not price in this underestimation in his bid is awarded in the auction. That is, if bidders do not adapt their bids accordingly in such a case with a high common value cost component, it is very likely that those bidders that underestimate the value of the common cost component most submit the lowest bids (based on their estimation) and are awarded. Furthermore, there is a high probability that this estimation was lower than the actual costs and thus those bidders may have a very low or even negative profit from realizing the project. The risk obviously increases if there are two common value components.[2] The risk of non-realization of the projects is therefore enlarged as awarded bidders are not willing to face a loss in the case of unfavorable development of the market conditions. Both awarded bidders in the first offshore wind auction in Germany already notified that they will make the final investment decision not before 2021 [10].

Furthermore, there is the economic risk of inefficiency. Efficiency means that those bidders are awarded in the auction that have the lowest electricity generation costs. However, in this auction efficiency is not ensured. For example, two bidders participate in the auction with one project each, where both projects have the same size ($s_1 = s_2$) and network connection point. If Bidder 1 has project costs of $c_1 = 4.7$ ct/kWh and estimates the average market price at $e_1 = 4.5$ ct/kWh, he will bid at least $b_1 \geq 4.7$ ct/kWh to receive a positive profit. On the other hand, if Bidder 2 has costs of $c_2 = 5.1$ ct/kWh but estimates a market price of $e_2 = 5.5$ ct/kWh, it is reasonable for Bidder 2 to bid $b_2 = 0.0$ ct/kWh. In this case, Bidder 2 has higher costs than Bidder 1 ($c_2 > c_1$), but would be awarded in the auction leading to inefficiency.

Another problem of this auction design is a possible accumulation of bids with 0.0 ct/kWh. In this scenario, the projects would be awarded according to their minimum bid quantity s_i^{min}, which represents the minimum capacity down to which the bid price b_i is valid, starting with the smallest one. This could also lead to an inefficient outcome. This inefficiency becomes even more striking by considering the additional cluster restrictions in this auction. It is possible that a 0.0 ct/kWh bid is not awarded whereas in the same cluster a smaller project with a higher bid is awarded. In the concrete example of the recent auction it could have happened that the EnBW Project 'He Dreiht' would not have been awarded if another project in the same cluster with a smaller minimum bid quantity would have bidden 0.0 ct/kWh as well. If this project were to be smaller than the cluster restrictions, another project in this cluster could have been awarded with a bid higher than 0.0 ct/kWh, e.g. a project with a 5.5 ct/kWh bid. This means in the same cluster the same capacity as the EnBW project would have been awarded, but at higher costs. It is furthermore actually possible that in this auction more projects submitted bids of 0.0 ct/kWh but have not been awarded due to general and cluster volume restrictions.

From bidders' perspective, the criteria so that a 0.0 ct/kWh bid dominates higher bids may not be very realistic as the average electricity wholesale market price is fluctuating from month to month. Even if the average market price is estimated correctly, that is $P = e_i$, for some months it may be lower ($p^t < e_i$), and thus the bidding interval which is dominated by a 0.0 ct/kWh bid is smaller. In principle, the actual interval of dominated bids is $[0, \min_t(p^t)]$ which is much more complicated to predict than P. In other words, a bid that was higher than 0.0 ct/kWh but lower than the average market price $b_i \in (0, P]$ would have resulted in a higher profit

[2] Two common cost components increase the overall share of common costs for the bidders and thus the risk of the Winner's Curse.

IOP Conf. Series: Journal of Physics: Conf. Series **926** (2017) 012015 doi:10.1088/1742-6596/926/1/012015

$$\sum_{t=1}^{T} \max(b_i, p^t) \geq \sum_{t=1}^{T} p^t. \tag{3}$$

The same holds for the uncertain future development of the wholesale market price. This development is subject to different influencing factors and might increase less than expected or even decrease, at least for some of the 25 years of contract duration. A bid that was originally dominated by a 0.0 ct/kWh bid due to the expected value of the future market price would insure the bidder against such scenarios.

And finally, it is striking that not all the auctioned capacity (1550 MW) was allocated. Four projects with combined 1490 MW of capacity were awarded in the auction and thus 60 MW were not. As a result, a bidder could have been awarded for this 60 MW with the ceiling price of 12 ct/kWh if he would have had a lower bidding limit of 60 MW or less. The fact that no bidder embraced this opportunity is even more surprising as the participants had the chance to submit two bids ranging over a installed capacity interval. Especially for smaller projects in the Baltic sea, this would have been a great chance.

4. Conclusion

The latest results from the first German offshore wind energy auction show that the prices for renewable energies are developing faster than expected and that there is a future need to advance the auction design. The main goal of such adaptions should be to obtain efficient outcomes and still incentivize further cost reductions. That is, auctioneers should focus on reducing the possibility of inefficient outcomes and at the same time to mitigate the risk of the Winner's Curse. Depending on the auctioned technology, there are different possibilities to achieve this goal through changes in the auction design, for example an adaption in the remuneration type to the respective market environment.

In the long-term it may even be possible that renewable energy sources do not need further support in the next decade, a result not foreseeable in past years. For a successful transition to such a scenario, it is important to not preclude possible market participants through an unfavorable auction design, e.g. not award bidders with 0.0 ct/kWh bids. Any future auction design has to keep the possibility in mind that bidders might be able to generate electricity at costs at or below the average market price. Nevertheless, the generated supply of renewable energy sources will be variable depending on the current conditions.

However, the bidders must adapt to the new situation also and carefully consider the risks and benefits of future projects and bidding strategies. Companies participating in auctions for offshore wind energy support have to consider the game theoretic nature of auctions when bidding. Their behavior and bidding strategy will influence the other players' future behavior in the market and a favorable short-term result may not be beneficial in the long-term. Therefore, it is vital to understand the auction design and its underlying structures so that the bidders can avoid risks and exploit opportunities.

References

[1] Deutscher Bundestag 2016 Gesetz zur Einführung von Ausschreibungen für Strom aus erneuerbaren Energien und zu weiteren Änderungen des Rechts der erneuerbaren Energien
[2] European Commission 2014 Guidelines on State aid for environmental protection and energy 2014-2020 (2014/C 200/01)
[3] Reed S 2017 Germany Strikes Offshore Wind Deals, Subsidy Not Included

[4] Bundesnetzagentur 2017 WindSeeG - 1. Ausschreibung für bestehende Projekte nach 26 WindSeeG Tech. rep. Bundesnetzagentur fir Elektrizität, Gas, Telekommunikation, Post und Eisenbahnen
[5] Kagel J H and Levin D 1986 *The American Economic Review* **76** 894–920 ISSN 00028282
[6] Thaler R H 1988 *The Journal of Economic Perspectives* **2** 191–202 ISSN 08953309
[7] Kreiss J, Ehrhart K M and Haufe M C 2017 *Energy Policy* **101** 512 – 520 ISSN 0301-4215
[8] Klessmann C, Wigand F, Tiedemann S, Gephart M *et al.* 2015 Designing renewable energy tenders for Germany
[9] Department for Business, Energy & Industrial Strategy 2017 FiT Contract for Difference Standard Terms and Conditions
[10] DONG Energy 2017 DONG Energy awarded three German offshore wind projects